MW00800715

TURTLES OF THE SOUTHEAST

Turtles

OF THE SOUTHEAST

by Kurt Buhlmann, Tracey Tuberville, and Whit Gibbons

The University of Georgia Press   Athens and London

Athens, Georgia 30602

Designed by Mindy Basinger Hill

Set in 10/15 Scala

Printed and bound by Four Colour Imports

The paper in this book meets the guidelines for permanence and durability of the Committee on Production Guidelines for Book Longevity of the Council on Library Resources.

Printed in China

12 11 10 09 08 P 5 4 3 2 1

Library of Congress Cataloging-in-Publication Data

Buhlmann, Kurt.

Turtles of the southeast / by Kurt Buhlmann, Tracey Tuberville, and Whit Gibbons.

p. cm.—(A Wormsloe Foundation nature book)

Includes bibliographical references and index.

ISBN-13: 978-0-8203-2902-4 (pbk. : alk. paper)

ISBN-10: 0-8203-2902-9 (pbk. : alk. paper)

1. Turtles—Southern States. I. Tuberville, Tracey. II. Gibbons, J. Whitfield, 1939– III. Title. IV. Series.

QL666.C5 B86 2008

597.920975—dc22 2007038567

British Library Cataloging-in-Publication Data available

Contents

SPECIES ACCOUNTS

PEOPLE AND TURTLES

TURTLES OF THE SOUTHEAST

Basking freshwater turtles (opposite) are a common sight in the Southeast.

Marine turtles ply the ocean waters near coastal southeastern states, and females come ashore to nest.

All About Turtles

WHY TURTLES?

People like turtles. Almost no one is afraid of them, and many people find them oddly appealing. The southeastern United States with its diversity of natural habitats offers anyone with an interest in nature endless opportunities to observe a variety of turtles as common components of the native wildlife scene. Anyone venturing near or on our lakes and rivers has seen painted turtles, sliders, or cooters basking on logs. Motorists have rescued thousands of box turtles from sure death on highways by assisting them to the other side. Countless people remember being fascinated as a child by a colorful spotted turtle or baby red-eared slider. And many an angler has had a line broken by the anticipated fish that turned out to be a large snapping turtle. Turtles are all around us in the Southeast.

One goal of this book is to teach people how to recognize and distinguish the many different species of turtles native to the region. An equally important purpose is to instill a sense of value and appreciation for these fascinating animals and an understanding of their role and significance in natural habitats. Many species are in serious trouble because of habitat degradation. A variety of environmental problems have led to their decline not only in our region but worldwide. In this book we suggest conservation strategies that could reduce the problems faced by turtles of the Southeast. The first step in promoting turtle conservation is to educate the public. Effective conservation strategies require an appreciation for a distinctive group of animals and familiarity with the various species in terms of appearances, geographic ranges, habitat associations, behaviors, and lifestyles. In this book, we hope to take this first step and a few beyond.

The gopher tortoise (*Gopherus polyphemus*, top) is one of several turtles endemic to the Southeast. The closely related desert tortoise (*Gopherus agassizii*, bottom) is the gopher tortoise's counterpart in the Southwest.

DEFINING THE SOUTHEAST

The Southeast has been defined in terms of its culture, its political boundaries, and its biology. For biological books that focus on a particular group of plants or animals, physiographic region (see map on p. 236), climate, and habitat are perhaps the most important factors. The distribution of habitats has the greatest effect on the distribution of turtles across the landscape, but for practical reasons we have used state boundaries to define the Southeast as follows: Alabama, Arkansas, Florida, Georgia, Kentucky, Louisiana, Mississippi, North Carolina, South Carolina, Tennessee, and Virginia.

NAMING TURTLES

Taxonomy is the science of classifying organisms. Taxonomists try to classify and name animals and plants in a way that indicates ancestral relationships (*phylogeny*), both among species within a group and between one group and others. Closely related species are grouped within a genus (plural = *genera*); closely related genera fall within a family. The selection of genus and species names follows a strict set of rules recognized by taxonomists around the world.

A typical scientific name includes the *genus* name (e.g., *Gopherus*), which is always capitalized, and a second name called the specific epithet (e.g., *polyphemus*), which is always in lowercase letters; both names are always in italics. Thus, the proper scientific name of the gopher tortoise is *Gopherus polyphemus*. This name indicates the species' close relationship to the desert tortoise, *Gopherus agassizii*. Both species are placed in the family Testudinidae, which also includes land tortoises of other genera, such as the giant tortoises of the Galápagos Islands in the genus *Geochelone*. The scientific name frequently describes a physical trait that is characteristic of the species or reveals an ecological or behavioral peculiarity. The gopher tortoise, for example, is named after Polyphemus, the Cyclops of Greek mythology who lived in a cave. Thus *polyphemus* refers to the gopher tortoise's proclivity for burrowing and inhabiting dark tunnels.

Regionally recognizable races may be further classified as subspecies, which adds a third element to the scientific name. For example, common mud turtles (*Kinosternon subrubrum*) are called *Kinosternon subrubrum subrubrum* in most of their eastern geographic range, but those found in southern Florida are considered a different subspecies, *Kinosternon subrubrum steindachneri,* because of their distinctive appearance. Likewise, mud turtles in the western part of the range have bright yellow stripes on the head and have been named *Kinosternon subrubrum hippocrepis.*

A common question concerns the distinction between turtles, tortoises, and terrapins. Which is what? All are turtles, whether they are land dwelling, seafaring, or residents of rivers or ponds. Typically a tortoise is a turtle that lives on land and belongs to the well-defined family (which includes the gopher tortoise) with elephantine hind feet and a domed shell. In the United States, the name "terrapin" refers to the diamondback terrapin of the Atlantic and Gulf estuaries and salt marshes. "Terrapin" is the Algonquin name for "edible turtle," which referred to any freshwater species. European settlers adopted the name for the diamondback terrapin based on the Algonquin definition.

The common names of turtle species do not follow the rules established for scientific nomenclature, and there is no universally accepted list of common names for amphibians and reptiles. Common names of turtles are typically based on what the people of a region have traditionally called them, and therefore often vary across the range of a species. For example, herpetologists (scientists who study reptiles and amphibians) usually call the turtles known scientifically as *Sternotherus odoratus* "musk turtles," but the same animals are called "stinkpots" or "stinking jims" in some rural areas of the Southeast. Other local names for turtles include "cooter," from the African word "*kula*" (turtle), and "slider," a term

A stinkpot or musk turtle.

likely given to describe the way many hard-shelled turtles "slide" off river banks, rocks, and logs. Forced standardization of common names seems a completely unnecessary exercise for professional herpetologists, who have enough difficulty agreeing on which scientific names to apply to species. Our approach in this book is to use the common names by which most people in the Southeast refer to particular species of turtles.

Because scientific understanding of relationships among species changes with new research, taxonomy is a volatile science. Originally, turtles and other species were classified according to similarities and differences in their appearance (i.e., morphology), but molecular genetics has given herpetologists a new perspective. The continuing advancement of molecular techniques allows us to detect smaller and smaller differences that may not always be ecologically important in distinguishing whether different populations are distinct species. Consequently, turtle taxonomists sometimes debate the proper interpretation and use of such information.

Taxonomists also sometimes disagree about the relative importance of different traits and what they reveal about the early relationships among turtle species, and not all taxonomists agree when scientific names are changed. And although herpetologists agree that all turtles are more closely related to each other than to any other group of animals, they are less certain about which other animal groups are turtles' closest relatives.

Because many of the proposed classifications are currently in dispute, we have generally used the traditional nomenclature along with a short note indicating changes that have been suggested. Our goal is to make certain that the reader knows what turtle we are referring to, no matter what scientific name is ultimately used.

Slider turtles may have been named for the way they "slide" from logs and stream banks.

TURTLE DIVERSITY

Turtles have a spectacular body design that has remained relatively unchanged for millions of years. The oldest known turtle fossil is 210 million years old and is from the Triassic period—long before the most famous of the giant dinosaurs walked the earth and before the earliest known snakes.

Turtles have adapted to many habitats. They exist in the world's oceans, in deserts, in tropical rainforests, and in myriad freshwater ecosystems—including swamps, ponds, and rivers. Over the millennia, turtles have made only slight modifications to their original body plan, and these reflect adaptations turtles have evolved to take advantage of different habitats. Marine turtles have flippers, for example, while tortoises possess elephant-like hind limbs that support their heavy bodies on land.

According to some estimates, there are 14 families and 316 species of turtles worldwide. Seven families are represented in the Southeast, and they include 42 species, or 13 percent of the world's total. The family Testudinidae (true land tortoises) has only one representative (gopher tortoise) in the Southeast. Kinosternidae (mud and musk turtles) is represented by 6 species, and the family Chelydridae (snapping turtles) by 2 species. There are 3 species of Trionychidae, the softshell turtles. The family Emydidae has by far the highest biodiversity, with 25 species occurring in the Southeast, including several found nowhere else in the world. Five of the world's 7 marine turtle species frequent southeastern coastal waters, some even nesting on the beaches. The family Cheloniidae includes the 4 southeastern species of hard-shelled marine turtles; the only species in the family Dermochelyidae is the leatherback, the world's largest turtle.

The Southeast is comparable to two other known areas of great turtle diversity: Southeast Asia and eastern India/Bangladesh. Of the 56 species of turtles known in the continental United States and its coastal waters, 75 percent are found in the Southeast, with Mississippi and Alabama tied for the most species found in any state at 30. The greatest concentration of species (18) is found in the lower Mobile Bay region of those two states.

The Southeast was probably a refuge for many northern species during glacial periods when North America was buried under ice from Ohio northward. These species have recolonized northern areas since the glaciers receded, but the number of species is far fewer in the north than in the Southeast. The Southeast's diversity of habitats, mild climate, and geologic history have combined to make the region rich in turtles and other biodiversity.

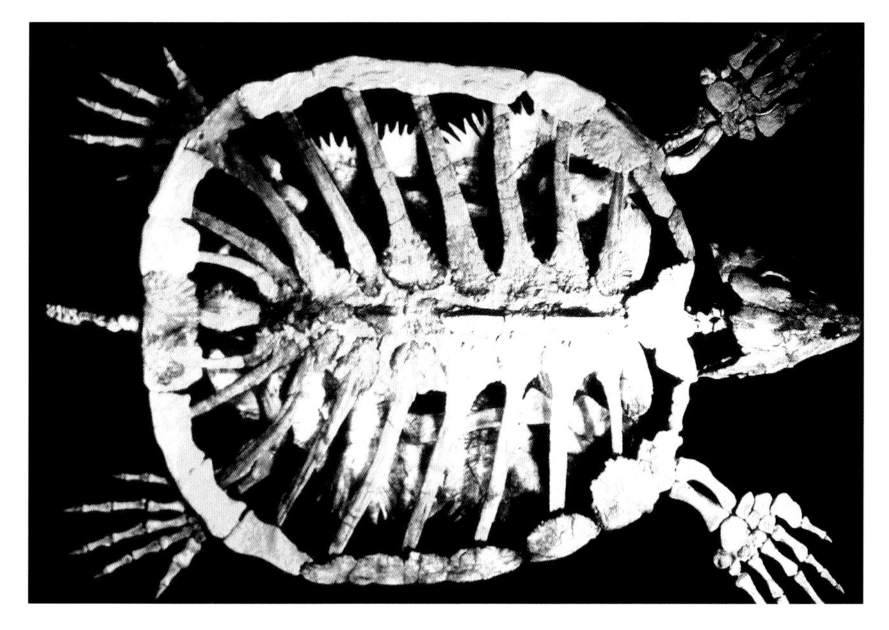

The bones in a softshell turtle's shell are greatly reduced.

General Biology of Turtles

MORPHOLOGY

Turtles have been traditionally classified as reptiles, along with lizards, snakes, and crocodilians. In fact, however, turtles come from an ancient lineage and may actually be no more closely related to other reptiles than they are to birds.

What makes a turtle a turtle? The most distinctive trait of a turtle is its shell, a protective plate of armor above and below. A turtle's shell is permanently attached to the rest of its body, and, cartoon renditions aside, turtles cannot crawl out of their shells. Turtles have many features that are characteristic of most other reptiles—they lay eggs and have scales, four limbs, and feet with claws—but they are the only reptiles with shells. And, like birds, turtles are distinct from most other vertebrates in having no teeth with which to grab prey or chew food. A few mammals, such as the armadillo, have shells, as do most mollusks, but the resemblance is superficial. The development and structure of those animals' shells differ greatly from the process and product in turtles.

Although they are relatively well protected by their remarkable shells, turtles of the Southeast have adopted a wide variety of defensive behaviors to adapt to their environments. Gopher tortoises burrow deep beneath

Did you know?

Although some species of snakes and lizards give birth to live young, all turtles lay eggs.

the ground, creating subterranean sanctuaries roomy enough for their high-domed shells. Flattened musk turtles have a hard, flat shell ideal for safely sequestering themselves in underwater rock crevices. Box turtles are the paragon of armored defense, being able to pull their head and limbs completely inside a hard, hinged shell that can be tightly shut. Softshell turtles have leathery, flattened shells and reduced bone structure that might make them more vulnerable to predators and other sources of mortality if it were not for the huge webbed feet that propel them rapidly through open water.

The extensive webbing on the feet of softshell turtles makes them fast swimmers.

Most turtles move overland with plodding resolve, as slow and persistent as the tortoise that outraced the hare in Aesop's fable; cooters and softshells, however, can travel at a surprisingly fast gait. Some freshwater and ocean species can swim faster than a person, although not as fast as a fish. A few southeastern turtles are wholly terrestrial, about half are almost exclusively aquatic, and the remainder divide their time between land and water. Life under these differing conditions requires different means of locomotion.

The anatomy of a turtle's feet indicates its habitat and customary mode of travel. Turtles that live in water most of their lives typically have webbed feet, with the amount of webbing reflecting the proportion of time spent in water. Sea turtles, which never leave the ocean except to lay eggs on a beach, are the extreme and have true flippers on both front and back feet. Softshell turtles seldom venture beyond the edge of a freshwater lake, river, or stream and have fully webbed feet designed to move them rapidly through the aquatic environment. Like the other aquatic and semiaquatic hard-shelled turtles, softshells have claws and digits on each foot that are useful for climbing onto logs for basking, walking along the bottom of a lake or river, and tearing apart prey or carrion. The greatest departure in footwear among southeastern turtles is seen in the gopher tortoises. With their hind legs built like an elephant's and front legs that look like flattened spades, gopher tortoises are designed both for walking on land and for burrowing under it.

Although all turtles have shells, there is quite a bit of variety in shell structure. Some species have thick, high-domed shells; others have shells ranging from steeply peaked to rounded. Some have shells that are flat and

The scutes are still attached to some of the bones of this turtle shell (top).

A turtle cannot be removed from its shell because the backbone is fused to the shell (bottom).

leathery. Others have shells with ridges down the center and on the sides or ornate structures along the top. But all turtles have a shell with a carapace above, a plastron below, and a bridge that connects the two. The margins of the carapace range from smooth and uniform in some species to slightly serrated or jagged in others.

Aside from skeletal features such as front and hind limb girdles being on the inside of the rib cage, the internal anatomy of turtles is similar to that of other vertebrates. All turtles have a heart, liver, pancreas, and spleen, as well as paired kidneys, lungs, and reproductive organs (ovaries or testes). In addition to using lungs for breathing air, some aquatic turtles get supplemental oxygen while underwater by means of special tissues in the throat and cloaca that extract oxygen from water.

The turtle digestive tract consists of an esophagus, stomach, and small and large intestines. The separate urinary system carries wastes from the kidneys to a large bladder.

THE SCUTES OF THE PLASTRON

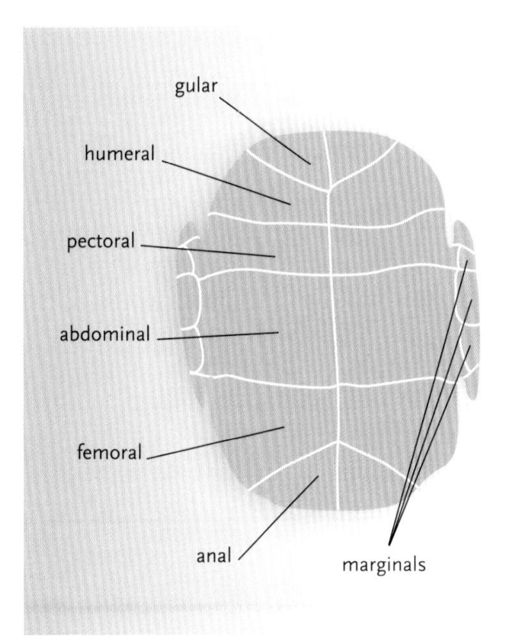

THE SCUTES OF THE CARAPACE

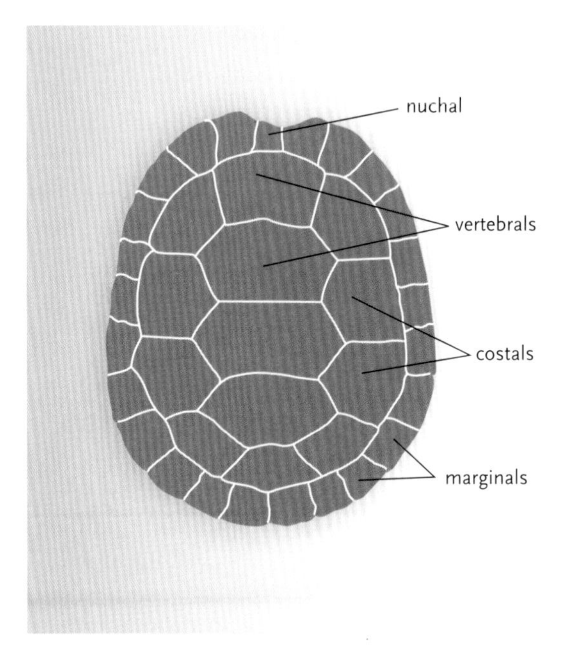

Wastes pass outside through the cloaca, a chamber into which the intestinal, urinary, and genital tracts all empty. The cloacal opening is visible on the underside of the tail of all turtles and usually extends just beyond the edge of the carapace.

Turtles have reproductive organs with the same basic design and function as those of mammals and birds. Males have a single copulatory organ, a penis, which is usually hidden inside the base of the tail. As a result, the tails of adult male turtles are characteristically longer and thicker at the base than the tails of females or juvenile males. During mating, the male turtle inserts his penis into the female's cloaca to fertilize the unshelled eggs she contains. The fertilized eggs are shelled in the oviduct, where they remain until they are ready to be laid. Female turtles of several species can store viable sperm in their oviducts for years. Because they also have the ability to store sperm from several matings, more than one male may father the eggs within a single clutch.

Turtles' basic senses of seeing, hearing, smelling, tasting, and feeling are the same as those of other vertebrates. Turtles have no ear openings but do have a tympanum and both middle and inner ears. They can presumably hear airborne sounds, although not very well. Aquatic turtles are sensitive to vibrations in the water, and gopher tortoises can detect vibrations transmitted through the soil, making it difficult for large predators or people to approach them unaware. Whether turtles have strong senses of smell and taste is not clear, although they apparently can detect bait in aquatic traps by the smell. All turtles see well at close distances, and some, such as cooters and map turtles, see well at great distances. All aquatic turtles presumably have good vision underwater.

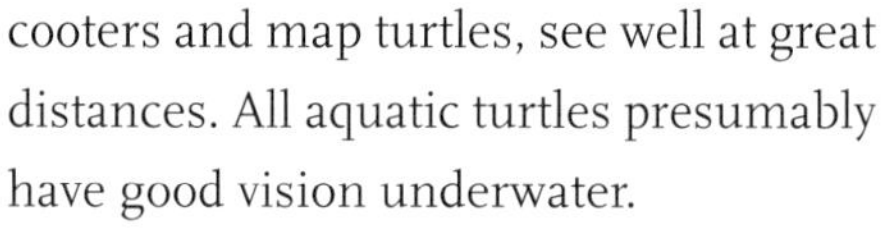

A long thick tail and concave plastron distinguish the male (left) from the female (right) in some turtle species.

A log covered in basking painted turtles is a common sight in some ponds in the Southeast.

ACTIVITY

Most people's outdoor experiences with turtles involve seeing them basking on logs or rocks along a river or pond, or finding them crossing a road. Some turtles bask openly, often in great numbers. Sliders, cooters, and painted turtles are especially prevalent on basking sites in spring and summer, but may also sun themselves on warm days in fall or winter. A careful observer traveling on a river where map turtles live is likely to see young and adult turtles basking on exposed stumps and trees that have fallen into the water. Snapping turtles and softshells also bask occasionally, usually at the water's edge. Musk turtles are more likely to bask in shaded sites than in open sun; usually they prefer the forks of small trees, sometimes up to 6 feet or more above the water! Mud turtles, on the other hand, almost never bask.

Softshells may bask in shallow warm water while hiding in the sand.

A turtle crossing a road usually has a purpose. Most turtles that live in isolated wetlands mate in late winter and early spring, and males move from one body of water to another in search of females, often crossing roads in the process. Later in the season, especially in late spring and early summer, females leave the water to find nesting sites and frequently have to cross a road to reach a suitable site. Even the highly aquatic map turtles must leave the water to lay eggs and are likely to be seen on highways near rivers. Many of the turtles seen on and alongside highways are females on

Aquatic turtles are sometimes found on land. They may travel between ponds, or they may aestivate on land when ponds are dry.

egg-laying excursions. A live female turtle sitting motionless by the roadside is probably laying eggs. Most turtles lay eggs in relatively open, sunlit areas, and therefore frequently choose road shoulders as nesting locations. Turtles of all ages and sizes commonly leave drying aquatic areas and travel overland in search of a more suitable habitat, often crossing highways along the way. In late summer and early spring, hatchling turtles travel from the nest to the nearest wetland. It takes a careful observer to spot one, though, even when it is crossing a road.

Aquatic turtles bask to raise their body temperatures enough to perform activities, such as feeding, digestion, and mating.

Turtles in cold regions hibernate from late fall to early spring, but many southeastern turtles become active for brief periods on warm, sunny days during the winter months. Sites chosen for winter dormancy vary. River turtles remain in the water, either hidden under banks, sitting exposed on the river bottom, or buried in mud or sand. Turtles that sit in cold water, especially when it is flowing, may obtain enough oxygen through their skin to maintain basic life support in their torpid state. Turtles that bury in mud often do not rely on obtaining oxygen from the water and must surface periodically to breathe air. Mud turtles and male chicken turtles in isolated wetlands hibernate on land beneath the forest litter; most sliders and female chicken turtles remain in the wetland. Box turtles dig under leaf litter or soft soil, burying themselves a few inches beneath the surface. Gopher tortoises spend cold days of fall, winter, and spring deep inside their burrows.

Although less predictable than the winter season, regional droughts can be a challenge to turtles. River turtles are less likely to be affected by major droughts because they simply retreat to the deepest remaining water. Small lakes, ponds, and other wetlands can dry up completely, however, leaving the turtles to take their chances in different ways. Sliders living in a drying lake, for example, can detect another body of water more than a quarter mile away and will move overland to it. The sensory mechanism they use remains a mystery. Mud turtles leaving a drying water body burrow on land nearby and become dormant, aestivating during the summer much as they hibernate during the winter. Because like all reptiles they can reduce their metabolic rate, turtles can sometimes survive simply by becoming inactive during stressful environmental times.

Some turtles migrate annually, the ultimate excursions being made by sea turtles between feeding sites and nesting beaches that may be hundreds of miles apart. Freshwater turtles leave drying wetlands in the fall, hiber-

Fires are a natural occurrence in southeastern pine forests, and box turtles occasionally get singed if they don't bury themselves deeply enough in the soil.

nate on land, and then return when the spring rains refill the wetlands. Movement among habitat patches is also common, although the reasons are not always apparent. For example, some gopher tortoises move several miles to another sandhill area while others of the same sex and size remain at the colony's original site. Likewise, many pond-dwelling turtles travel overland for miles to another wetland while others stay behind. Individual softshell turtles have been recaptured at locations several miles apart, and individual map turtles have been known to move more than 3 miles up and down rivers.

All species of southeastern turtles are active during the day, and a few aquatic species also prowl around in the water at night. All species except the sea turtles generally nest during the daylight hours; sea turtles nest almost exclusively at night. Snapping turtles sometimes lay eggs at night as well, and map turtles and wood turtles sometimes start nesting in early evening and finish after dark. Aside from mating activities, a typical day for a freshwater turtle during favorable weather consists of periods of basking, foraging, and sitting dormant in a protected area such as in heavy aquatic vegetation, under the bank, or concealed beneath mud or sand. Many turtle species are especially active after a warm rain.

Did you know?

Freshwater turtles respond to drought in several ways. Some species nestle into the mud of the drying wetland, others bury themselves in the surrounding terrestrial habitat, and some travel overland in search of a more permanent aquatic habitat.

FOOD AND FEEDING

As a group, turtles are dietary generalists. Some species are carnivores, eating mostly other animals; some are herbivores, eating mostly plant material; but most are omnivores, eating both plant and animal material, both intentionally and incidentally. The powerful beak of carnivorous turtles such as alligator snappers is more than adequate to kill or subdue live prey.

Many, perhaps most, turtles also scavenge on dead fish and other animals. For example, a gopher tortoise, a species that feeds almost exclusively on grasses and leafy plants, was observed in southern Georgia feeding on a road-killed armadillo!

The biting parts of a turtle's toothless mouth are sharp ridges of bone that form the upper and lower jaws. Some jaws are smooth and razorlike; others are serrated like a steak knife. Turtles are able to use these sharp edges to shear off bite-sized chunks, whether of plant or animal, and swallow them without further chewing. The upper jaw of some species has a curved beak with a pointed tip. Some map turtles have broad, thick jaw surfaces for crushing snails and mussels that they pick up from the river bottom.

Herpetologists place reptiles in two prey-hunting categories: sit-and-wait ambush predators and wide-ranging foragers. Most turtles qualify for the second category as they actively move around in search of prey or plant foods. Foraging turtles find most of their food by sight, although aquatic turtles presumably locate dead fish by smell. Box and wood turtles walk through the forest in search of mushrooms, edible plants, berries, and small animals. Chicken turtles search for aquatic insects and crayfish in wetlands by poking their long neck into crevices and between patches of aquatic vegetation. And mud and musk turtles travel slowly along the bottoms of ponds, ready to grab any small animals they encounter; some species "feel" along in the dark with their chin barbels.

A few strictly aquatic turtles take advantage of camouflage and powerful jaws to surprise unsuspecting prey. Alligator snappers sit on the bottom, open mouthed, and use their pink, wiggling tongue as a lure. Fish in search of an easy meal become one instead. Spiny softshell turtles, with sandy-colored shells that conceal them beneath a layer of creek sand, are in an ideal setting to extend their long neck in a rapid strike to capture small fish

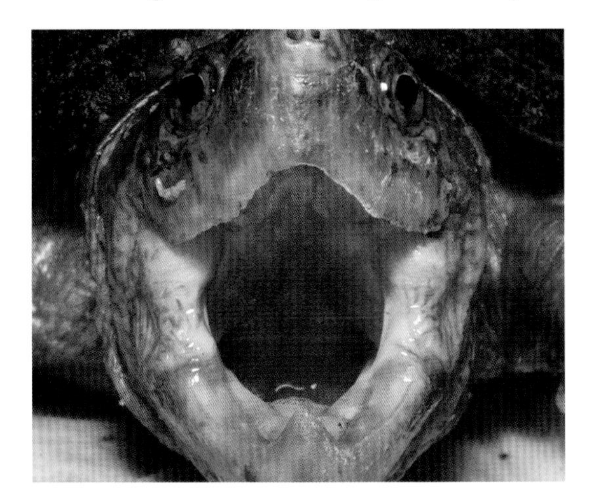

Carnivorous turtles (left) have sharp and straight-edged jaws. Herbivorous turtles (right) have serrated jaws that allow them to shred plants.

Box turtles are omnivorous, eating both plant and animal material.

or invertebrates that swim within range. Wood turtles sometimes use an unusual strategy called "worm stomping," tapping the soil with their feet and plastron in spots where earthworms are likely to be underground and eating the worms that come to the surface. Why the stomping brings the worms to the surface is unknown.

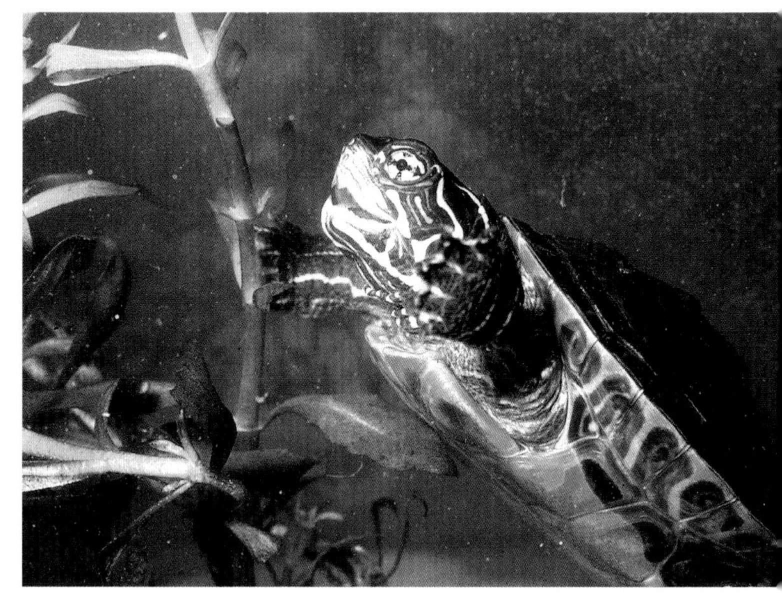

A herbivorous red-bellied cooter eating aquatic vegetation.

Almost every type of plant or small animal is the food of one type of turtle or another. Map turtles are noted for specializing on mollusks. Diamondback terrapins eat a variety of marine animals, including blue crabs, shrimp, and periwinkle snails. River cooters may use their serrated jaws to shred blades of aquatic plants waving in a river current; and box turtles readily eat mushrooms, blackberries, and wild strawberries. An inattentive small snake, salamander, frog, or tadpole is likely to fall prey to almost any hungry box turtle it encounters. Snapping turtles have gained notoriety for occasionally eating baby ducks and geese. But snappers are not picky eaters; their overall diet encompasses small mammals, snails, frogs, algae, duckweed, and lily pads. They will even eat small turtles, including their own species.

The diet of most turtles varies seasonally according to prey availability, but it also changes with age and size. Many species are carnivorous during their juvenile growth years and become herbivorous as adults. Insects and other animals contain more of the calcium young turtles need to develop their shells. For example, baby cooters and sliders eat mostly insects and other aquatic animals, whereas adults can subsist strictly on plant material, although they will eat fish, meat, or invertebrates when the opportunity presents itself.

The head size of some species of carnivorous turtles is related to their diet. The head of an adult female diamondback terrapin or Barbour's map turtle, for example, is disproportionately larger than the head of a male of the same body size. As a result, females eat larger prey, and the two sexes may eat entirely different types of animals. In contrast, the heads of older male loggerhead musk turtles and the Florida subspecies of the common mud turtle are noticeably larger than those of females. Whether males and females of these species have different diets has not been determined.

Turtles can go without eating for days, weeks, or even months under natural drought conditions or in response to changing seasons. During periods when wetlands partially or completely dry up, aquatic turtles burrow under banks; in mud; or beneath leaves, logs, or soil on land, where they remain dormant and may not eat for long periods. Most southeastern turtles eat little or nothing during cold periods in late fall and winter, which may be prolonged in northern areas.

REPRODUCTION

In many southeastern species, the males have long foreclaws.

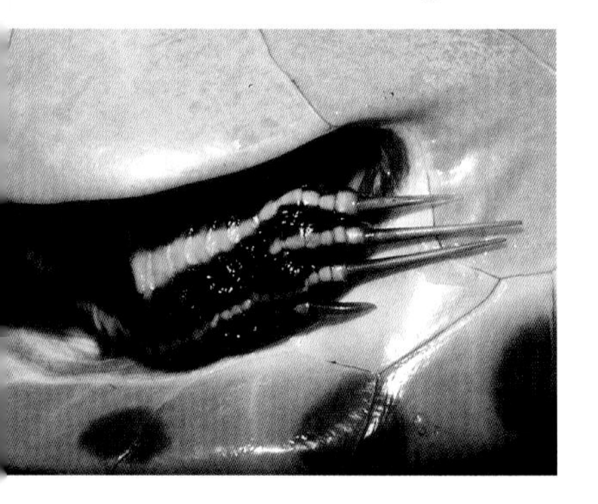

The general reproductive behavior of turtles is similar to that of other reptiles, birds, and mammals. Male turtles seek out adult females and mate with them by means of internal fertilization. A few turtles, such as cooters, sliders, painted turtles, and some map turtles, have intricate courtship behaviors in which the male vibrates his long front claws in front of the female. Complete courtship has not been reported for most aquatic turtles because of the difficulty of observing them in their natural habitat, and complex mating behavior may be more common among aquatic turtles than is currently known. Intricate courtship displays including vigorous head bobbing, biting, or both have been observed in the terrestrial gopher tortoise and some map turtles. Males of some species (e.g., gopher tor-

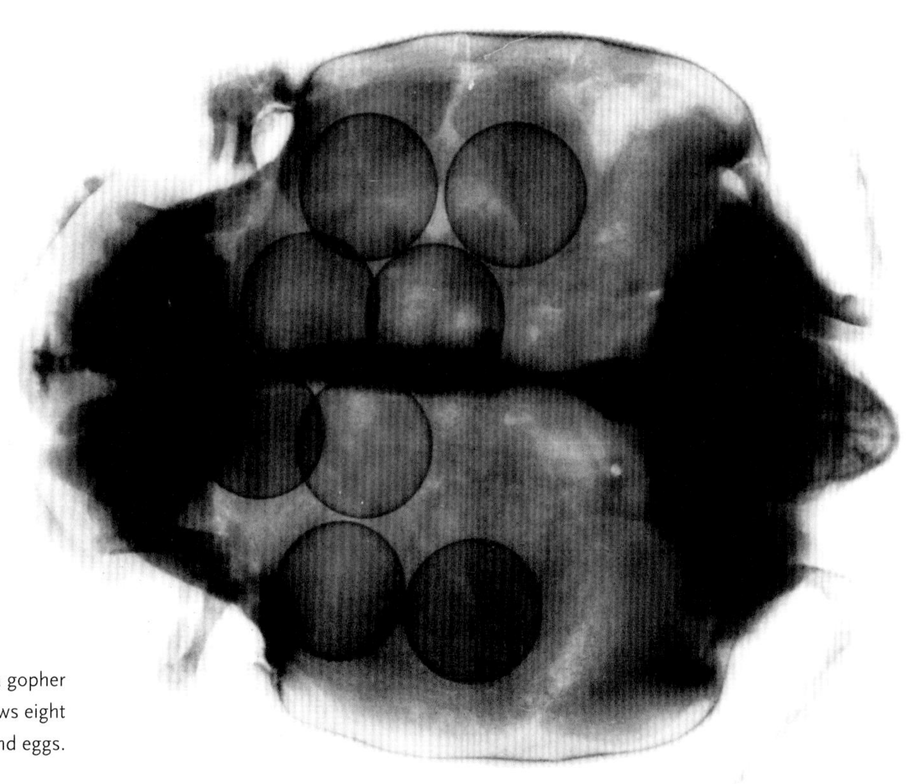

This X-ray of a gopher tortoise shows eight round eggs.

A flashlight illuminates a turtle egg. The reddish mark is the embryo. Turtle eggs cannot be rolled once they begin to develop, or else the embryos will suffocate.

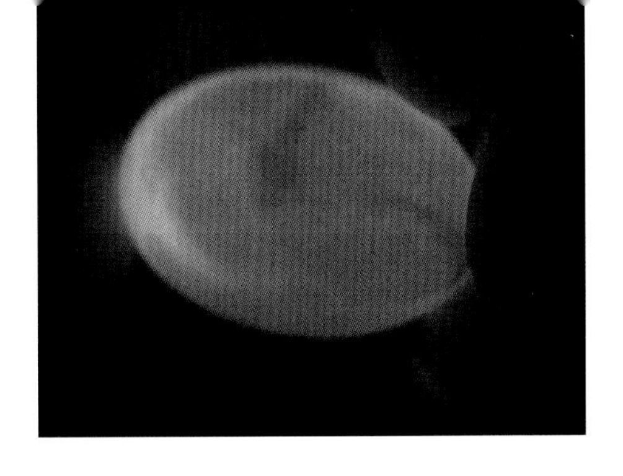

On either side of the main nest cavity, a river or pond cooter sometimes digs a satellite nest that may serve as a decoy for predators. One egg is usually deposited in each satellite nest.

Bog turtles will nest in sphagnum or grasses (below).

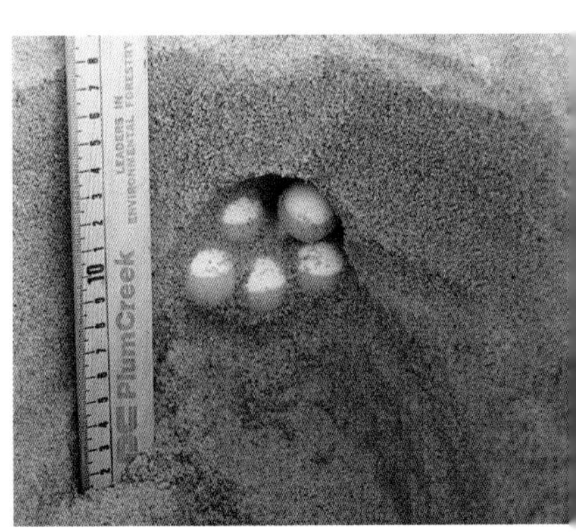

A cross section of a softshell turtle nest shows that the eggs are deposited in a flask-shaped chamber and not encased in sand.

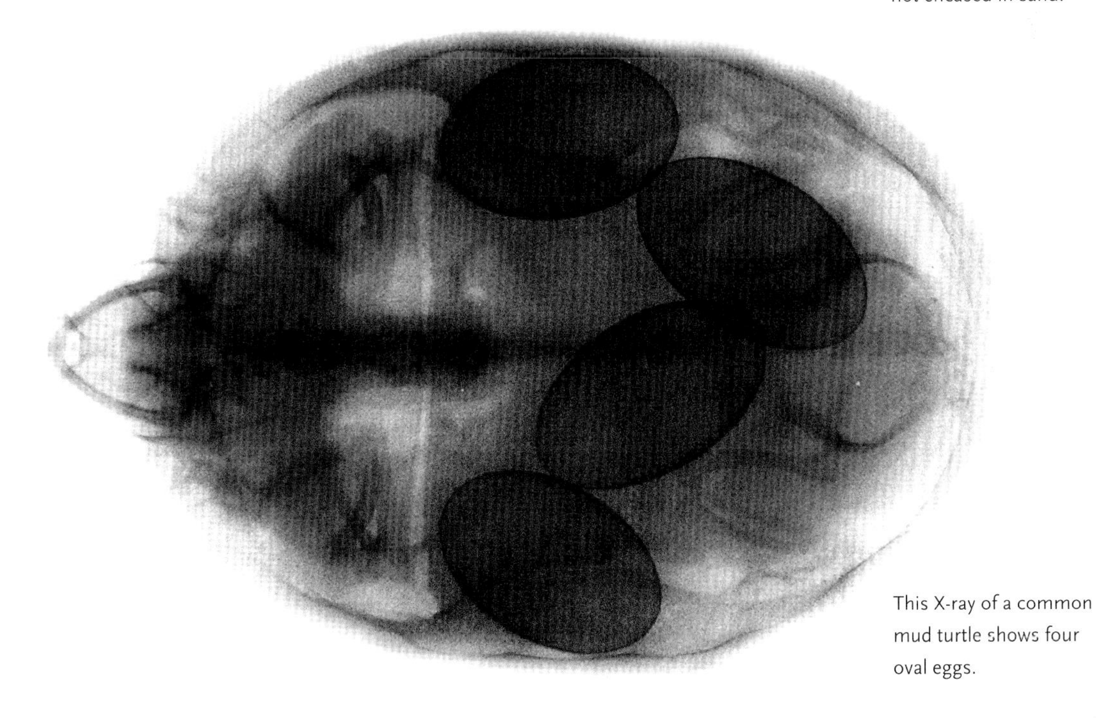

This X-ray of a common mud turtle shows four oval eggs.

A female slider turtle (top left) and a female snapping turtle (top right) digging a nest and depositing eggs.

A painted turtle urinates while digging a nest. The urine helps soften clay soils.

toises and snapping turtles) engage in combat by ramming or physically overpowering each other.

During mating itself, the male turtle normally crawls atop the shell of the female (in water or on land, depending on the species) and wraps his tail (which is usually longer than the female's) around her tail. When their cloacas are in contact, the male's penis can be inserted into the female. After fertilization, the male and female disengage and go their separate ways.

Most southeastern turtles mate primarily in early spring and in the fall, but a few mate throughout their summer activity period. All turtles bury their eggs in sand, soil, or decaying vegetation, usually in nests the female digs with her hind legs. All turtle eggs are white but vary in shell shape and texture. The eggs of softshells, both species of snapping turtles, gopher tortoises, and sea turtles are round; those of all other species are oval or elliptical. Shell texture varies from leathery in some species to hard and brittle in others.

Did you know?

During the nesting process, a female will use her rear legs to guide the eggs into the nest as they are laid.

After depositing the eggs and covering the nest, the mother turtle typically departs and provides no further care for the eggs or the hatchlings. A few species nest in somewhat peculiar locations that may provide some protection to the nest after the female has left it. For example, Florida red-bellied cooters sometimes nest in alligator mounds, and gopher tortoises sometimes place their nests well inside their burrows. Snapping turtles in Michigan sometimes nest in ant mounds, but this behavior has not been observed in the Southeast.

Although most southeastern turtles lay their eggs from late spring into summer, a few species nest on warm days in fall and winter. In effect, one turtle species or another may be nesting in almost every month of the year. The onset and length of the nesting period for species with broad geographic distributions may vary significantly from region to region as a result of local climate.

Eggs laid in spring generally hatch after an incubation period of 2–3 months. Tem-

Most hatchling turtles do not emerge from the nest until the yolk sac is completely absorbed.

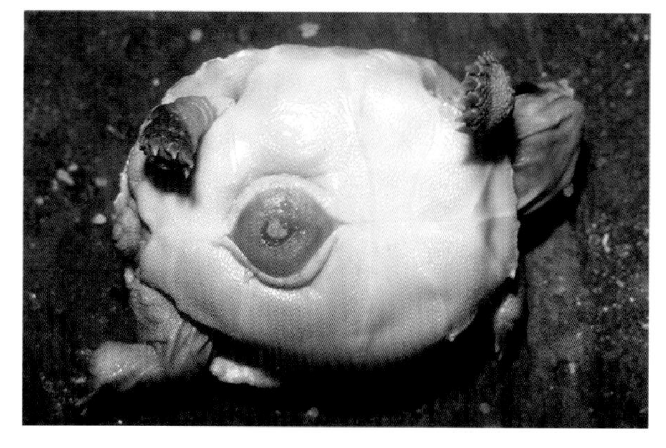

The yolk scar is visible for some time after the yolk sac has been absorbed.

A diamondback terrapin hatchling emerges from its egg.

The caruncle, which is used to slice open the egg shell, is visible on this hatchling.

perature affects not only the incubation time of turtle eggs, which develop more rapidly at warmer nest temperatures, but also, for most species, the sex of the hatchlings. Although the process is complex and varies among species, eggs incubated at warmer temperatures tend to produce females, and those incubated at cooler temperatures tend to produce males. The sex of softshell turtles is determined genetically.

The eggs of most aquatic species hatch in mid to late summer, but the hatchlings remain underground in the nest chamber through the fall and winter, emerging in the spring of the following year. Even hatchlings that emerge from the nest in the fall will spend a few days inside the nest after hatching, absorbing the large yolk sac that is attached to their plastron and provides fuel during their first days outside the nest. Some hatchlings have a structure called a caruncle, or egg tooth, on the end of their snout with which they slit open the eggshell. The caruncle falls off soon after the turtle is free of the egg.

Did you know?

An Australian turtle has been discovered that lays eggs underwater, but the eggs start to develop only after the waters recede.

DEFENSE

The persistence of turtles in the fossil record for millions of years underscores the success of their basic body plan. The combination of carapace and plastron creates a shelter for the protection of limbs, head, and vulnerable soft body parts. The bony shell of a turtle is derived from the ribs, which have flattened and fused together to form a continuous bony plate. The plastron is derived from the front of the rib cage and is comparable to the sternum or breastplate in other vertebrates. The backbone of a turtle is fused to the center of the carapace. Because the neck and tail are extensions of the backbone, as in humans and other vertebrates, a turtle cannot be removed from its shell. Of particular interest and continuing scientific marvel is the ability of turtles to actually withdraw their limbs inside their shell, which would be equivalent to a human withdrawing arms and legs inside the rib cage!

Turtles have modified the basic shell plan in a variety of ways that can be observed in southeastern turtles. Box turtles have a hinge on the front third of the plastron that allows them to pull the plastron up tight against the carapace, a very useful survival strategy since box turtles share the forest floor with a host of predators such as raccoons and foxes. The semiaquatic mud turtles also have hinges that allow them to partially close their shell, both protecting them from predation and also helping to avoid water loss while on land. Slider turtles have a thick, bony shell that may help protect them from the alligators with which they often share the water. Cooters and map turtles have slightly thinner shells than sliders but are fast and nimble swimmers.

Some species have additional defensive mechanisms and rely less on the shell for protection. Snapping turtles have traded in shell defense for attitude and aggressiveness. Their plastron is very small, and snappers cannot pull their heads or limbs very far into the shell, but they have large

Concealment is a protective strategy employed by many turtles.

The gopher tortoise digs a burrow for protection from predators and the elements.

A box turtle can close up its plastron tight against the carapace.

claws and powerful jaws with which they readily scratch and bite. Softshell turtles have traded armor for speed. They have much less bone in their shell than the hard-shelled turtles, and instead have leathery skin covering their thin, nonconnected ribs. However, softshell turtles are agile and can swim incredibly fast. They also spend much of their time hiding, buried underwater on shallow sandbars in the rivers they inhabit. Some turtles emit a noxious smell when disturbed. Musk turtles—or stinkpots, as they are so deservedly called—will emit a pungent substance from the joints where the plastron meets the carapace when they are disturbed. Presumably this deters some predators, although raccoons are known to eat musk turtles. Gopher tortoises have a strong shell but also make their own refuge: burrows averaging 15 feet long and 6 feet deep in the sandy pine savannas in which they live. Finally, some turtles simply hide to avoid being discovered. Chicken turtles and mud turtles bury themselves in the leaves and humus of the forest floor when their wetland home is dry. Flattened musk turtles wedge themselves into narrow rock crevices underwater.

PREDATORS

Despite their impressive body armor, turtles in the Southeast have many natural predators as well as some introduced ones. Even humans eat them. In fact, turtles are a regular part of the human diet in many regions of the world, with the food market demand in Asia affecting millions of turtles each year, including some from the Southeast.

Fire ants are an exotic species that kill many hatchling turtles shortly after they hatch but before they can emerge from the nest.

The egg and hatchling stages of every turtle's life are times of great vulnerability to predation. Terrestrial predators, including raccoons, gray foxes, skunks, opossums, and crows, are all known to uncover freshly dug nests and eat the eggs. Raccoons may follow a female turtle's tracks and scent to find her nest. Nests that survive undetected for a few nights have a dramatically greater likelihood of survival. Some predators kill nesting female turtles, which are more vulnerable when they are on land. The density of some predators, especially raccoons, increases with increased human development, both because they have subsidized food sources (i.e., garbage) and because their own natural predators (wolves and cougars) are usually absent.

Turtle eggs and hatchlings are vulnerable to introduced fire ants, an invasive species from South America that is now well established in the Southeast. Fire ants kill hatchling turtles in the nest, including those that remain in the nest over the winter. The full impact of fire ants on turtle populations is unknown because their predation on underground nests is not visible from the surface, but these ants occur widely in the Southeast.

Once aquatic hatchling turtles reach the water, wading birds (e.g., herons and egrets), large fish (e.g., gar and bowfin), and bullfrogs are waiting to snap them up. Hatchlings of terrestrial species must contend with large predatory snakes, such as the indigo snake and coachwhip. Adult freshwater turtles are not immune to predation either. American alligators can crack the shells of most turtles. Only the largest and heaviest species, such as alligator snapping turtles, yellow-bellied sliders, and Florida red-bellied turtles, are relatively safe from them.

Alligators (left) can prey on turtles, such as the softshell pictured here.The high domed shell of pond cooters (below) can provide protection from an alligator's jaws.

Raccoons, (top) a native predator of turtles and turtle nests, are abnormally abundant in many areas due to the lack of larger predators and the presence of human-provided sources of food.

Raccoons often follow the scent of a female turtle's trail from the water's edge to the nest. Here (bottom) a raccoon's tracks flank a turtle's.

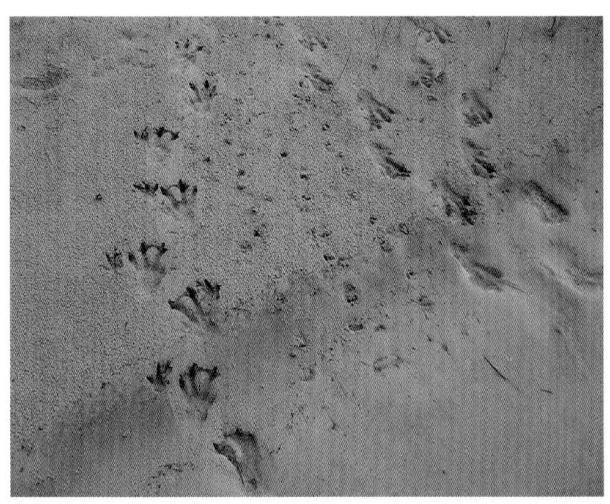

River otters can inflict serious damage on freshwater turtle populations. A single river otter that ventures into a seasonal wetland can eliminate a major portion of the resident chicken turtles and musk turtles. Otters even attack snapping turtles during winter, when the turtles are inactive.

Feral dogs and coyotes also pose hazards for adult turtles. Box turtles in suburban areas are often found with severely chewed shells, indicating that they survived a bout with someone's unsupervised pet. Coyotes and feral hogs will dig up and consume juvenile gopher tortoises and semiaquatic turtles hiding under leaf litter during aestivation.

Finally, although the shell of the turtle has proven a worthy defense against natural predators for millions of years, it is no match for the automobile. Among the greatest sources of mortality to southeastern turtles are cars and trucks that hit individuals attempting to cross roads. Most of the turtles killed in this way are females searching for a place to dig a nest. The chronic loss of these animals poses a threat to turtle populations throughout the Southeast.

Did you know?

After reaching maturity, a female turtle continues to lay eggs throughout her life, which may last decades.

AGE AND LONGEVITY

In addition to their protective shell and the fact that the incubation temperature determines the sex of individuals of many species, turtles have two more traits that set them apart from almost all other animals: the age of individuals can be estimated with accuracy for several years after hatching, and individuals live a long time relative to other animal species. Many species of tortoises have been reported to live more than 50 years in captivity, alligator snapping turtles have lived for more than 70 years, and at least one Blanding's turtle (a species found in the Midwest and Northeast) laid eggs at an age of more than 80 years.

Most longevity records are based on captive animals. Studies of natural populations rarely produce such information. However, a turtle's age can sometimes be determined by an examination of annuli—rings on the scutes that are formed during periods of slow growth, such as winter—in much the same way that trees can be aged with tree rings. And, like the use of the technique with trees, the rings and the spaces between them tell how many winters a turtle has lived through; in other words, how old it is. The rings are not visible on all species of turtles (softshells and marine turtles have no apparent rings) and begin to fade and disappear completely on older individuals of most species. Growth in juvenile turtles is relatively rapid, but it slows once an animal reaches reproductive maturity. Thus, the age of an individual can often be reliably estimated up to maturity, but after that the separation between

This juvenile slider turtle is in its fourth year, as indicated by the number of annuli.

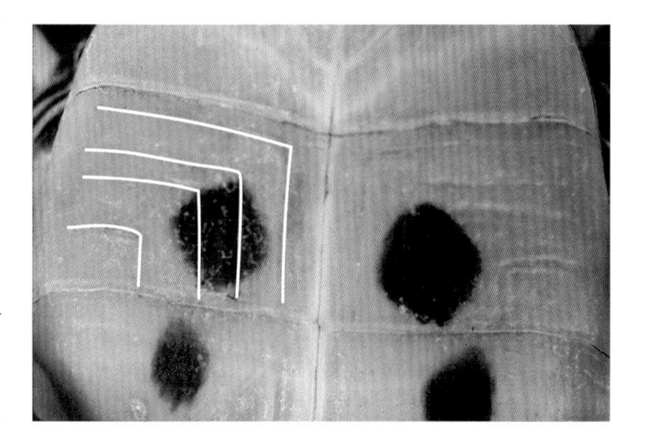

The number of annuli indicates that this juvenile wood turtle is in its fourth year.

After a turtle reaches maturity, growth slows or stops and the turtle's age cannot be accurately determined by counting annuli.

The innermost ring indicates the size of the scute at hatching.

Some species display growth rings for many years.

rings is not detectable. The accuracy of this method is high for some turtle species; it is sometimes possible to count 15–20 rings on old box turtles, gopher tortoises, and wood turtles, for example. Slider, painted, and map turtles often have 6–10 clear rings. When the mark-release-recapture and annuli aging techniques are used together, the age of much older turtles can be determined if they were captured and accurately aged when they were young and still growing. Thus, a 7-year-old slider or painted turtle that is recaptured 23 years later is without doubt a 30-year-old turtle. Such records have been collected from several long-term turtle studies in the Southeast such as the Savannah River Ecology Lab.

MATURITY

All turtles take much longer to reach maturity than most other animals. White-tailed deer and cottontail rabbits, for instance, can reach sexual maturity in less than a year. Hatchling songbirds return the next spring to make nests of their own. Opossums, raccoons, gray foxes, and skunks are all adults in the year following their birth. Early maturity is also generally coupled with a shorter lifespan, and many of these animals live less than 10 years.

Turtles, in contrast, are long-lived species with delayed maturity. Males of most freshwater turtles, including map turtles, painted turtles, and sliders, take a minimum of 4 or 5 years to reach maturity. As many as 7–10 years seem to be the average for males of bog, spotted, wood, and box turtles. Gopher tortoise males require 15–18 years to become adults. Females of all

Turtles are long-lived animals. Here, Whit Gibbons holds a mud turtle that he marked before coauthor Tracey Tuberville was born.

turtle species require even more time than males to mature. Chicken turtle females mature in 5 years, and 7–10 years is the norm for Barbour's map turtles, sliders, painted turtles, and cooters. Some species, including ornate box turtles, wood turtles, and common snappers, may not be able to mate until they are 12–15 years old. Gopher tortoise females may take as long as 18–20 years. The disparity between the sexes is most striking in Ernst's map turtle; males may mature at age 3, but females are not mature until age 14. Marine turtles vary in age at maturity, with hawksbills and Kemp's ridleys requiring 4–6 years, leatherbacks 12–13 years, and loggerheads and green sea turtles perhaps even longer. Most, if not all, female turtles lay eggs throughout their entire lives, meaning that 80-year-old females may still be digging nests.

Turtles take a long time to reach maturity, and most juveniles do not make it that far, but once a turtle reaches adulthood and has a strong, hard shell, its probability of surviving from one year to the next is high. Because all female turtles presumably lay eggs throughout their entire lives but only a small number of their offspring survive to maturity, persistence of turtle populations depends on high survivorship of adults. Populations that experience artificially high rates of adult mortality—whether through harvesting for food, road mortality, or some other reason—can dwindle rapidly and may never recover.

Giant tortoises from the Galápagos Islands and from the Indian Ocean's Aldabra group are known to live more than 100 years.

This diamondback terrapin is the only southeastern turtle species that is resident in brackish water habitats.

Turtle Habitats in the Southeast

Turtles of the Southeast live in all sorts of habitats. Some habitats support more species, or more individuals of a single species, than others. Some turtle species have precise habitat needs and are found only in a limited area; others can use a variety of habitats. The location of a habitat on the landscape relative to other habitats is also a factor in determining which species are likely to be present. Below we describe the most common habitats used by turtles of the Southeast.

Most people have their first experience with a turtle on land. Although all southeastern turtles nest on land, and many may aestivate on land or travel overland between aquatic habitats, only the gopher tortoise and perhaps the ornate box turtle live their lives exclusively in terrestrial habitats. Most southeastern turtles are associated with freshwater systems—marshes and swamps, seasonal ponds and lakes, rivers, and reservoirs. Exceptions include the diamondback terrapin, a denizen of salt marshes, and the sea turtles, which live in the open ocean and in coastal estuaries.

A pond cooter in a Carolina bay wetland.

Because sunlight does not reach the floor of densely planted pine forests (left), food production for box turtles and gopher tortoises is limited. A savanna-like longleaf pine forest (right) is ideal habitat for gopher tortoises.

TERRESTRIAL HABITATS

Pine Forests

The only resident turtle of the pine forests is the gopher tortoise. The Coastal Plain pine forests often contain sandy, well-drained soils in which gopher tortoises construct their extensive burrows. Historically, this fire-maintained system resembled a savanna or even a prairie with sparsely scattered longleaf pine trees. It provided extensive open areas for basking and nesting and a diverse understory of plants for foraging. Other terrestrial turtles, such as the common and ornate box turtles, may pass through pine forests, and some aquatic turtles, such as common mud turtles and chicken turtles, may aestivate or hibernate in pine forests adjacent to their wetlands.

Old fields and farms (left) often contain common or ornate box turtle populations. A prairie-like wetland (right) is ideal for Gulf Coast box turtles.

Mixed Hardwood Forests

Mixed hardwood forests occur throughout the Southeast. These forests, with their thick humus and leaf litter, provide shady, moist retreats for box turtles and aestivating aquatic turtles from adjacent wetlands. Hardwood forests are important foraging areas for box turtles and wood turtles, which rely on the fruits, seeds, fungi, and invertebrates found there. Small openings in the canopy created by downed trees or other disturbances may also provide nesting sites.

Prairies and Old Fields

Natural prairies and agricultural fields occur within forested habitats and are found along rivers and near other aquatic habitats. Many turtle species nest in these open areas, and wood turtles and ornate box turtles frequently forage in these habitats.

Box turtles require hardwood forests.

FRESHWATER WETLAND HABITATS

"Wetland" is a term that is difficult to define. It is commonly used to identify just about any piece of land that is flooded with water either permanently or seasonally, or that just feels squishy underfoot. The word "wetland" contains three letters referring to water and four letters indicating land. Thus, wetlands can be considered the interface between terrestrial and aquatic habitats. The terrestrial habitat surrounding the water can be just as important as the water itself to the residents of wetland habitats. The diversity of wetland types is spectacular, and each is unique in some way; thus, categorizing them and the turtles that frequent them is rather difficult.

Yellow-bellied sliders, painted turtles, and snapping turtles often use beaver ponds.

Farm and Beaver Ponds

Beavers create ponds by damming small streams. As the beavers cut and eat shrubs and small trees, the remaining larger trees drown and eventually fall. These fallen logs provide basking structures for aquatic turtles such as painted turtles, northern red-bellied turtles, and common musk turtles. Box turtles that live in surrounding hardwood forests often soak at the edges of the water. Historically, the presence of beaver ponds likely allowed many semiaquatic turtles to disperse across the landscape by using these habitats as stepping-stones.

Farm ponds often have many of the same turtle species as beaver ponds. The species found in farm ponds will be determined by pond size, depth, the amount and diversity of shoreline, and submerged vegetation, as well as the presence and abundance of logs for basking. In the Piedmont and mountain regions, farm ponds are home to painted turtles and common snapping turtles. Ponds in the Coastal Plain typically contain sliders, pond cooters, Florida softshells, and common mud and musk turtles.

Painted turtles, pond cooters, sliders, and stinkpots inhabit ponds that retain water year-round.

Isolated Seasonal Wetlands

Isolated seasonal wetlands can vary tremendously in size, depth, how frequently they dry (if ever), and the type of vegetation in and along their margins. Typically such wetlands are not connected to streams and do not contain fish. They may be small sloughs with grassy bottoms that dry frequently each summer. Some are dominated by standing forests of bald cypress and black gum trees with the high-water mark evident several feet up on their trunks. A few (for example, Okefenokee Swamp in Georgia, Payne's Prairie in Florida, and Lake Waccamaw in North Carolina) are quite large. These wetlands are extremely productive habitats, providing a plethora of tadpoles, aquatic insects, crayfish, and aquatic plants on which turtles can feed. As a result of their inherent variability, isolated seasonal wetlands such as Carolina bays and limestone sinkholes will differ in the particular turtle species that occupy them. Included among the species that frequent these habitats are common mud turtles, striped mud turtles, chicken turtles, sliders, pond cooters, common snapping turtles, spotted turtles, and common musk turtles.

Seasonally filled wetlands are important habitats for spotted, mud, and chicken turtles. Such wetlands are often destroyed by human activities. Seasonal ponds such as this one change character with the seasons.

Isolated cypress ponds (left), found on the Coastal Plain, are home to chicken, spotted, and mud turtles. Carolina bays (middle) are unique egg-shaped isolated wetlands. Large wetlands and swamps (right) are home to Florida red-bellied cooters and pond cooters.

Wet mountain meadows are the only places where the secretive bog turtle resides in the Southeast.

Did you know?

Turtles can travel long distances between wetlands. Yellow-bellied sliders have been captured up to 5.5 miles overland from their original capture location.

Wet Meadows and Bogs

In the Blue Ridge Mountains, wet meadows and bogs exist where small streams meander through old pastures or the basins of abandoned beaver ponds. Such habitats are home to the secretive bog turtle and common box turtles. In the Coastal Plain, pitcher plant bogs are often frequented by box turtles, particularly the Gulf Coast subspecies.

RIVER SYSTEM HABITATS

Rivers and Reservoirs

Southeastern rivers have extremely variable characteristics that depend on the topography and geology of the region through which they pass. Rivers that originate in the mountainous limestone regions of Tennessee, Kentucky, and Alabama are among the most aquatically diverse in the world. Common map turtles, stripe-necked musk turtles, river cooters, spiny softshells, and the Cumberland subspecies of sliders all reside here. The unusual flattened musk turtle hides among underwater rocky crevices in the Black Warrior River of northern Alabama. The black tannin-stained rivers of the lower Coastal Plain have extensive sandbanks that provide critical nesting habitat for spiny and smooth softshells, alligator snapping turtles, and some map turtles. The numerous log snags found in most rivers offer excellent basking sites.

Several southeastern turtles are endemic to a particular river. For example, the ringed, yellow-blotched, Barbour's, Alabama, Gibbons', Ernst's, and Sabine map turtles are each restricted to one river drainage. Likewise, different parts of a single river drainage may have different turtle faunas; for example, some species are found only above or only below the Fall Line, where the Piedmont meets the Coastal Plain. The extensive ledges of weather resistant rock along the Fall Line provide abundant basking sites for turtles such as river cooters and may also be a barrier to movement upstream or downstream.

Few turtle species prefer the open water of the deep man-made reservoirs that now cover large sections of southeastern rivers. The shallow marshy edges of reservoirs often provide the same habitat structure and have the same turtle species as are found in small ponds, but the original riverine species are reduced or absent.

Map turtles, river cooters, and alligator snappers will nest on sandy beaches along Coastal Plain rivers (top). Mountain and Piedmont rivers (top middle) are home to stripe-necked and flattened musk turtles. Coastal Plain rivers along the Gulf Coast (bottom middle) are home to map turtles, alligator snappers, and razor-back and loggerhead musk turtles. Where southeastern rivers encounter the rocky Fall Line (bottom), river cooters are often abundant.

Loggerhead musk turtles, Florida red-bellied cooters, and Florida softshells are found in big springs (above). Wood, musk, and snapping turtles often occur in small streams (middle).

Floodplain Swamps and Oxbows

Most southeastern rivers are bordered by extensive floodplain swamps, especially in the Coastal Plain, where currents slow, sediments drop out, and river channels widen. Old river channels called oxbows create still backwater, almost pondlike areas. Floodplain swamps and oxbow lakes may experience cycles of drying and flooding but are often teeming with fish and riverine turtles, including sliders, pond cooters, spiny softshells, and alligator snapping turtles. Spotted turtles or striped mud turtles often live in the shallow upper reaches of the forested floodplain swamps.

Streams and Springs

Smaller streams that flow through forests and fields are likely to contain painted turtles, common snapping turtles, and common musk turtles. Wood turtles in Virginia also inhabit small streams. Some of the large Florida springs are the headwaters of small rivers, and their clear waters are home to Florida red-bellied cooters, Suwannee River cooters, and loggerhead musk turtles.

Floodplain swamps can be home to striped mud turtles, alligator snappers, and pond cooters.

BRACKISH WATER HABITATS

Salt Marshes

The extensive salt marshes of the Atlantic and Gulf coasts are the exclusive home of only one species of nonmarine turtle, the diamondback terrapin. Some of our marine turtles, notably loggerhead sea turtles, enter the estuaries in summer to feed on crabs and shellfish.

Salt marsh creeks are home to diamondback terrapins.

MARINE HABITATS

Open Ocean and Estuaries

Five species of marine turtles frequent our estuaries, including large tidal bays and the mouths of Atlantic and Gulf coast rivers. The Chesapeake Bay in Virginia is especially important as the summer feeding ground of juvenile Kemp's ridley sea turtles. Loggerhead sea turtles nest on the ocean beaches of all coastal Southeast states. Beaches along the coast of Florida and occasionally some other states provide nesting habitat for green and leatherback sea turtles.

Marine turtles and occasionally diamondback terrapins nest on ocean beaches of the U.S. Southeast.

A juvenile Florida softshell.

Species Accounts

ORGANIZATION AND ORDER OF SPECIES ACCOUNTS

The order of the species accounts is designed to help the reader become familiar with the species of turtles native to the Southeast. The species are presented in five major groups—terrestrial, semiaquatic, riverine, brackish water, and marine—according to the areas and habitats where they are most likely to be discovered.

Maps showing the species' distribution accompany each account. The larger map shows the species' range within the Southeast and illustrates state boundaries and, for riverine turtles, major river drainages. The smaller map shows the entire range of species, including the portions outside the Southeast. The maps are based on records from natural history museums around the country. We have indicated the natural geographic boundaries for species that are confined to specific river drainages or occur only in certain physiographic regions such as the Blue Ridge Mountains or Coastal Plain. Keep in mind that a specific habitat, such as a seasonal wetland, will not be found uniformly throughout a given species' range. Unfortunately, the contemporary distributions of some species may not be as extensive as the maps indicate because human development has eliminated large areas of some habitats.

TERRESTRIAL TURTLES include the species—such as the gopher tortoise and box turtle—that live all or most of their lives on land. Box turtles will occasionally visit the edges of wetlands or floodplain swamps.

SEMIAQUATIC TURTLES encompass a wide range of habitat use and behavior. They can be divided into two subcategories: (1) species that actually live in both aquatic and terrestrial habitats according to the season (for example, wood, mud, and chicken turtles) and spend part of the year in water and part of the year either foraging or aestivating in terrestrial habitats; and (2) primarily pond-dwelling species (for example, painted turtles, sliders, and pond cooters) that frequently traverse land between aquatic habitats but regard the terrestrial habitat only as a travel corridor.

BASIC FEATURES OF THE SPECIES ACCOUNTS

The carapace is brown to olive green. This individual is an adult female.

How do you identify a Sabine map turtle?

DISTINGUISHING CHARACTERS
Small spot behind each eye; many thin, faint stripes on neck; males with very narrow head; found only in the Sabine River drainage

AVERAGE SIZE

FEMALE
MALE
HATCHLING

CARAPACE SHAPE

FRONT VIEW

SIDE VIEW

TOP VIEW

Sabine Map Turtle *Graptemys sabinensis*

DESCRIPTION These small to medium-sized turtles are among the smallest of the map turtles. The carapace is brown to olive green, often with thin yellow concentric lines on each carapace scute. The keel on the carapace is prominent, especially on males, and the posterior of the shell is serrated. The plastron is basically plain white or cream colored, with darker markings in the seams. Numerous thin yellow neck stripes reach the back of the eye; a small spot is also present behind the eye. The females weigh three to four times more than the males but do not have a broad head. Males have a very narrow head and slightly elongated foreclaws.

VARIATION AND TAXONOMIC ISSUES Sabine map turtles are similar to Ouachita map turtles, but some turtle biologists recognize them as a distinct species. Sabine map turtles have a small roundish or oval spot behind each eye, and Ouachita map turtles have a large blotch.

WHAT DO THE HATCHLINGS LOOK LIKE? Sabine map turtle hatchlings are similar to the adults but have a nearly round shell. The carapace keel is more prominent and the carapace edges are serrated.

Males have a very narrow head.

178 • *Sabine Map Turtle*

RIVERINE TURTLES are almost always associated with permanent aquatic habitats such as rivers, streams, and impounded reservoirs, although some of the semiaquatic turtles—specifically sliders, pond cooters, musk turtles, and Florida softshells—will visit flowing streams and rivers as well.

BRACKISH WATER TURTLES are represented by the diamondback terrapin, the only species that lives exclusively in our coastal salt marshes.

MARINE TURTLES are species of the great oceans that may also frequent coastal estuaries.

To see the order of species accounts at a glance, consult the chart on page 235.

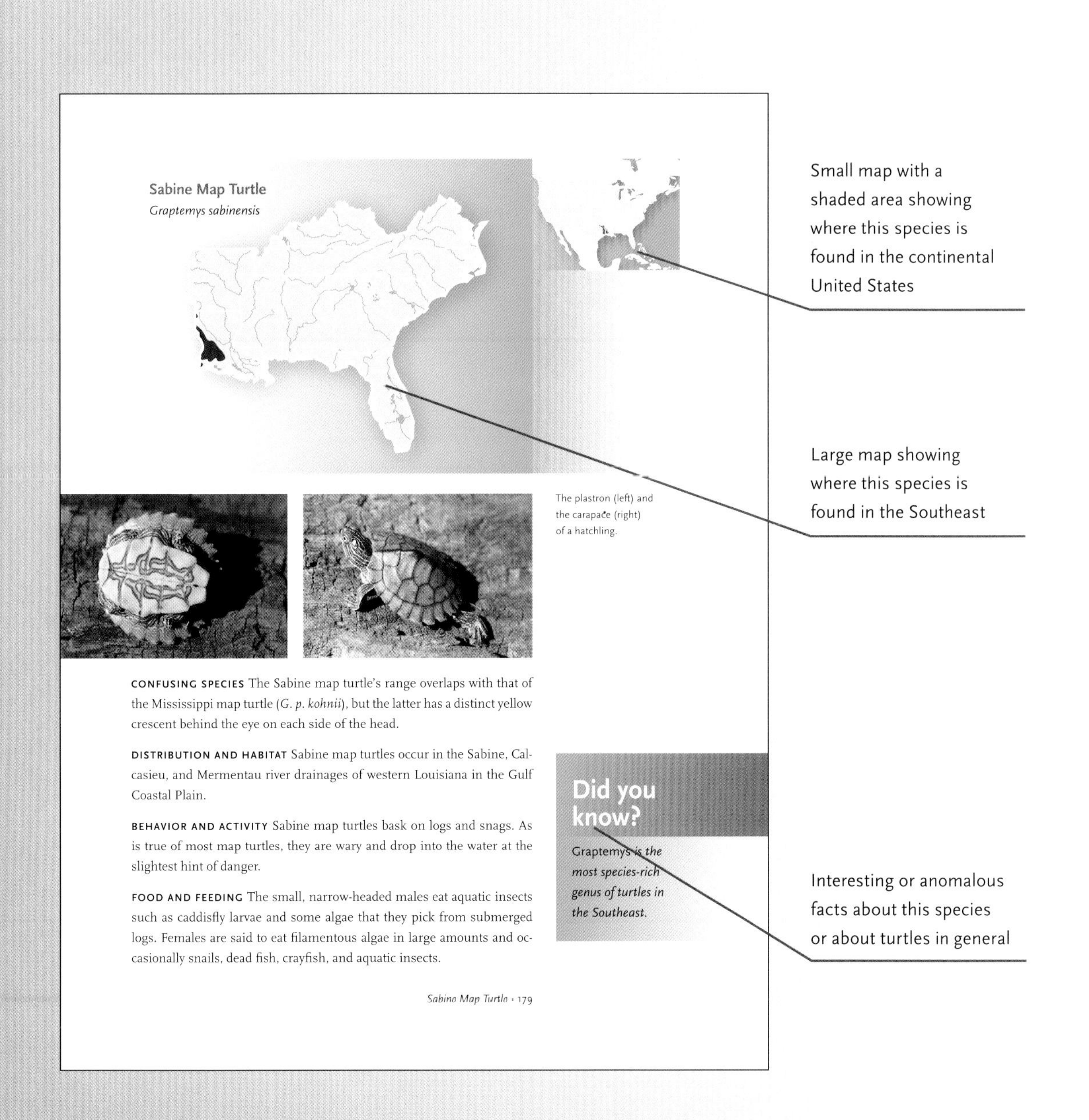

Sabine Map Turtle
Graptemys sabinensis

The plastron (left) and the carapace (right) of a hatchling.

CONFUSING SPECIES The Sabine map turtle's range overlaps with that of the Mississippi map turtle (*G. p. kohnii*), but the latter has a distinct yellow crescent behind the eye on each side of the head.

DISTRIBUTION AND HABITAT Sabine map turtles occur in the Sabine, Calcasieu, and Mermentau river drainages of western Louisiana in the Gulf Coastal Plain.

BEHAVIOR AND ACTIVITY Sabine map turtles bask on logs and snags. As is true of most map turtles, they are wary and drop into the water at the slightest hint of danger.

FOOD AND FEEDING The small, narrow-headed males eat aquatic insects such as caddisfly larvae and some algae that they pick from submerged logs. Females are said to eat filamentous algae in large amounts and occasionally snails, dead fish, crayfish, and aquatic insects.

Did you know?
Graptemys is the most species-rich genus of turtles in the Southeast.

Sabine Map Turtle · 179

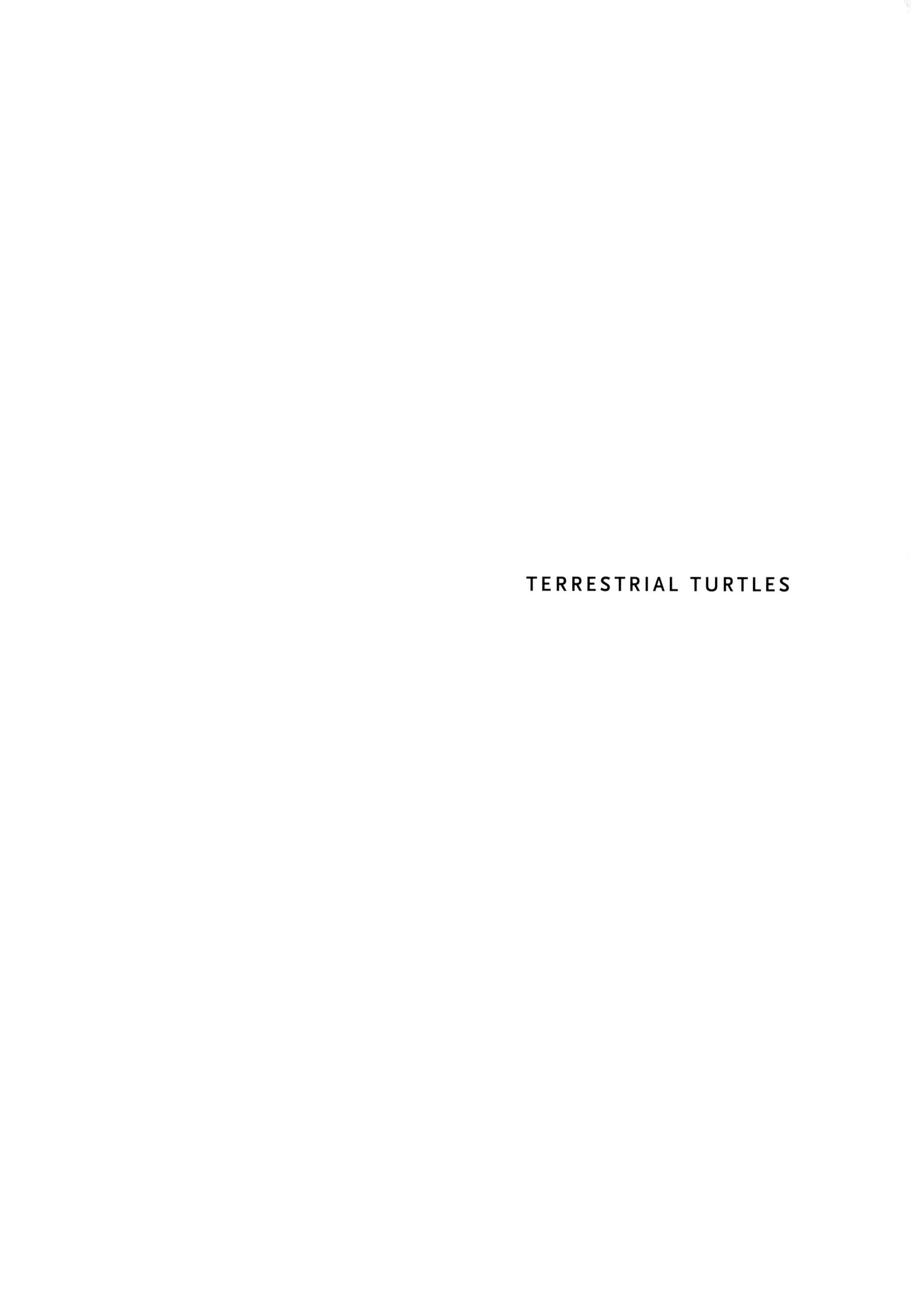

TERRESTRIAL TURTLES

A subadult with concentric annuli and sculptured shell.

Gopher Tortoise *Gopherus polyphemus*

DESCRIPTION The adult gopher tortoise has a domed shell that is unmarked and uniformly dark brown to grayish black. The plastron is lighter than the carapace. It has no hinge, but is deeply notched at the tail and is elongated at the front, forming a gular projection. The front limbs are wide, flat, armored with hardened scales, and end in large, flat nails that are used for digging. The hind legs resemble those of an elephant. Young tortoises (to about 12 years of age) have distinct growth rings, but the carapace becomes smooth in older individuals, erasing the rings. As males approach maturity, at around 8–9 inches, the plastron becomes concave and the gular projection lengthens. A pair of enlarged, wartlike glands can be found on the underside of the chin of both sexes, but the glands are more conspicuous in males than females, particularly during the breeding season.

VARIATION AND TAXONOMIC ISSUES No subspecies are recognized, and there is no regional variation in appearance.

Elephantine limbs with prominent scales and short claws characterize the gopher tortoise's rear foot.

How do you identify a gopher tortoise?

DISTINGUISHING CHARACTERS
Elephantine hind legs; nubby tail; wide, flattened nails on front feet; prominent gular; hardened scales on limbs; always on land

AVERAGE SIZE

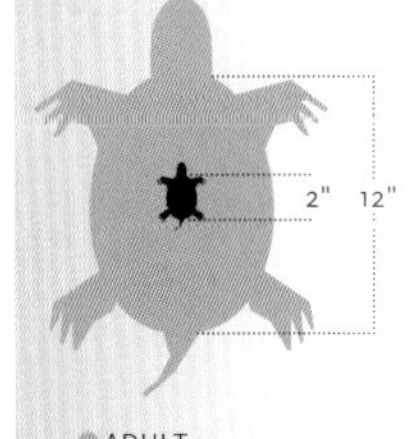

● ADULT
● HATCHLING

CARAPACE SHAPE

FRONT VIEW

SIDE VIEW

TOP VIEW

Hatchlings are brightly colored.

WHAT DO THE HATCHLINGS LOOK LIKE? Hatchlings are small, rounded, more colorful versions of the adults. The carapace is yellowish brown, and each scute usually has a bright yellow or orange center. The shell of young tortoises is pliable, but it becomes hardened with age.

CONFUSING SPECIES An adult gopher tortoise is unlikely to be confused with any other southeastern turtle. Young tortoises can be distinguished readily from box turtles by their elephantine rear legs and lack of a hinge on the plastron. Box turtles have clawed and webbed feet and a hinged plastron.

DISTRIBUTION AND HABITAT Gopher tortoises occur in the Coastal Plain from southern South Carolina south throughout Florida and west into southeastern Louisiana. Although they occupy areas with clay soils in some parts of their range, tortoises are more typically associated with deep, sandy soils in which burrows can be easily excavated. Gopher tortoises historically occurred in the savanna-like longleaf pine forests that once dominated the region but now occupy a wide variety of open habitats with abundant ground-cover vegetation that provides forage. They prefer frequently burned longleaf pine and scrub oak forests but will also use road

The gopher tortoise uses powerful front feet to dig burrows.

and utility rights-of-way, field edges, and fencerows. Because open areas are important for thermoregulation and nesting, tortoises avoid areas with thick, shrubby vegetation.

BEHAVIOR AND ACTIVITY The gopher tortoise is the only turtle in the Southeast that digs its own burrow. The width of the burrow typically indicates the length of the tortoise, as the animal must be able to turn around in it. The conspicuous underground retreats of adults may have an entrance more than a foot wide and a tunnel that averages 15 feet in length and angles down 6 feet below the surface. Even hatchling tortoises can dig their own burrows shortly after leaving the nest.

Individuals spend most of the time underground in their own burrow, safe from temperature extremes, predators, and the occasional fire, but will frequently visit tortoises residing in other burrows. Tortoises are most active from May through September but will emerge from their burrow whenever warm weather allows. In southern Florida, gopher tortoises may be active throughout the year.

As males approach maturity, the plastron becomes concave and the gular projection lengthens.

FOOD AND FEEDING Gopher tortoises are almost exclusively vegetarians, feeding on grasses and other low-growing herbaceous plants, and on fruits or berries when these are available. Although tortoises have been observed feeding on carrion, feces, and old

A gopher tortoise basks on its apron outside of its burrow in a longleaf pine–wiregrass savanna.

Gopher tortoises nest at the entrance to or even inside the burrow.

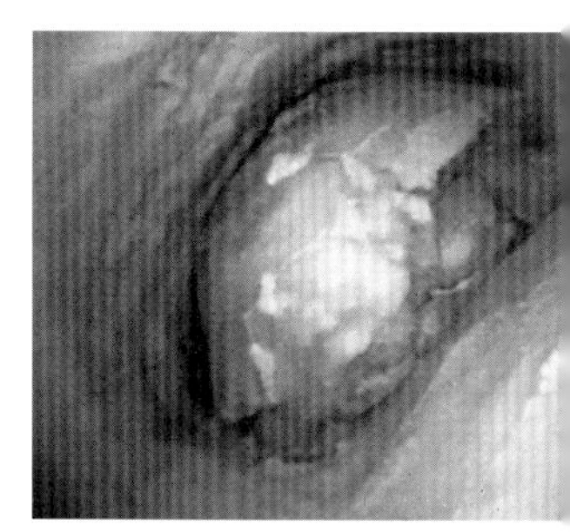

A gopher tortoise burrow is shaped like the tortoise—rounded on top and flat on the bottom.

The powerful front legs of a tortoise fling sand as it excavates its burrow.

bone fragments, they probably do so only opportunistically or when they are deficient in an important nutrient.

A hatchling gopher tortoise emerges from the egg.

REPRODUCTION Courtship and mating occur commonly in the fall and sometimes in spring, often directly outside the female's burrow. The female produces only one clutch per season, usually laying 3–9 (average = 6) ping-pong-shaped eggs in a nest just outside the burrow entrance or in the tunnel itself but sometimes in another open, sunny area. Eggs are laid in May–early July and hatch in late August–early October. Recently emerged hatchlings may seek temporary refuge under leaf litter, pine straw, or other ground cover, but each will dig its own burrow before winter arrives.

PREDATORS AND DEFENSE Eggs, hatchlings, and small juveniles with shells that are still soft and growing are eaten by raccoons, opossums, foxes, and skunks, and by large snakes such as coachwhips and indigo snakes. Introduced fire ants, which have spread throughout the tortoise's range, can be a threat to eggs and juveniles. Armadillos and feral pigs sometimes find nests while rooting for food. Adult tortoises are fairly well protected against most natural predators but may be vulnerable to large carnivores such as coyotes, bobcats, feral dogs, and large raccoons.

Did you know?

Eggs of most turtle species are elliptical or oblong, but several species in the Southeast (softshells, snapping turtles, gopher tortoises, and sea turtles) have round eggs that look like ping-pong balls.

CONSERVATION ISSUES The gopher tortoise is federally protected in Louisiana, Mississippi, and western Alabama and is a species of conservation concern throughout its range. Populations have declined throughout the range, and individual populations remain vulnerable to extirpation. Historical threats included consumption by humans and the gassing of tortoise burrows for collection of diamondback rattlesnakes, which use the burrows as retreats during winter and early spring. Because gopher tortoises are long-lived, take 15–20 years to reach maturity, and produce few young, many affected populations have not fully recovered from these historical impacts. Habitat destruction, fragmentation, and poor management (e.g., lack of prescribed fire, establishment of densely stocked pine plantations) of preferred habitats continue to exacerbate the species' decline. A contagious respiratory illness may have contributed to some population declines. Urbanization, particularly in Florida, has eliminated much of the gopher tortoise's habitat. As tortoises are forced to live in open roadside habitats, road mortality becomes an increasing threat. The survival of this species will depend on aggressive protection and intensive management of the remaining habitat patches and individual tortoise populations, including reintroductions into unoccupied sites.

Ornate box turtles have less color variation than common box turtles.

How do you identify an ornate box turtle?

DISTINGUISHING CHARACTERS
Can completely close up inside high-domed shell; radiating pattern of yellow and black lines on plastron and carapace; usually found on land

AVERAGE SIZE

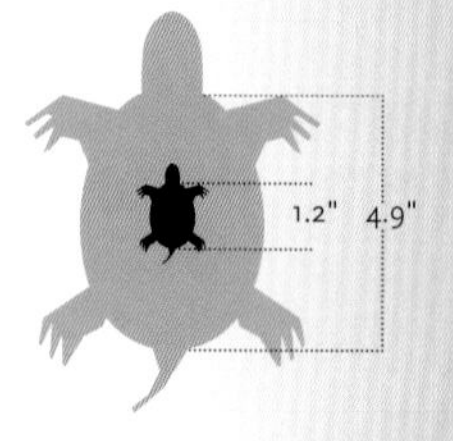

● ADULT
● HATCHLING

CARAPACE SHAPE

FRONT VIEW

SIDE VIEW

TOP VIEW

Ornate Box Turtle

Terrapene ornata

DESCRIPTION Ornate box turtles have the domed shell characteristic of most terrestrial turtles. The brown carapace has a broken or complete yellow stripe down the center, and each scute has yellow lines or stripes radiating from its center. The plastron is hinged in the front and is dark with yellow lines on all scutes. The head and limbs have yellow spots. The hind foot typically has four toes, and males have a curved inner toe on each hind foot that helps them clasp the female's shell during mating. Adult males have a slightly concave plastron and bright red eyes; adult females have a flat plastron and yellow to brown to maroon eyes.

The plastrons are patterned much like the carapaces.

GEOGRAPHIC VARIATION AND SUBSPECIES Of the two subspecies, only one (*T. o. ornata*) is found in the Southeast. No substantial geographic variation of this subspecies occurs in the southeastern portion of the range.

WHAT DO THE HATCHLINGS LOOK LIKE? Hatchlings have yellow markings around the margin of the shell and on each scute of the carapace, which has a bright stripe down the center. Hatchlings are more round than adults and the plastral hinge is not functional.

Hatchlings are brightly colored yellow and black.

CONFUSING SPECIES The only turtle in the Southeast that might be confused with an ornate box turtle is the three-toed box turtle, *T. carolina triunguis,* which has only three toes on each hind foot and a distinct vertebral keel, and does not have yellow stripes on the carapace. The Florida box turtle, *T. c. baurii,* has a similar pattern of radiating yellow and black stripes on the carapace, but not on the plastron.

DISTRIBUTION AND HABITAT In the Southeast, ornate box turtles reach the eastern extent of their range in the Ozark and Ouachita mountains and the Gulf Coastal Plain of Louisiana and Arkansas. They are primarily a species of western prairies and grasslands; they may be found in oak savannas but generally avoid closed-canopy hardwood forests. Ornate box turtles are more tolerant of dry, sandy conditions than are common box turtles.

BEHAVIOR AND ACTIVITY Ornate box turtles are fully active from March to November and are inactive in the winter months except during exceptionally warm spells. During periods of hot weather they are most active in

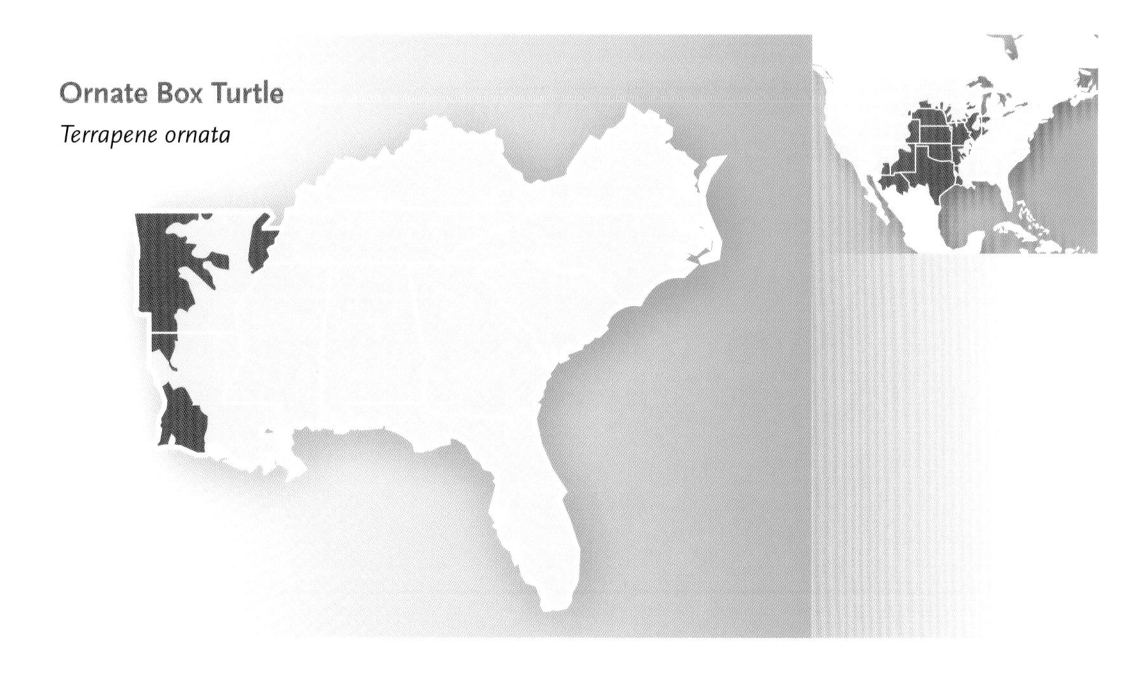

early morning and late afternoon. They characteristically bury themselves in the soil, at the base of plants, or under ground litter at night and during the hottest parts of the day or soak in standing water if it is available. They are especially active during and after rains. They travel overland for long distances in search of feeding or nesting sites and mates, and may migrate to and from hibernation sites and warm-weather activity areas.

FOOD AND FEEDING Ornate box turtles are active carnivores but will also eat berries and cactus and readily scavenge on dead animals. Among their known prey are eggs and young of ground-nesting birds, caterpillars, and especially dung beetles.

Did you know?

Turtles that spend periods of drought or hot weather on land will often return to the same location year after year to bury themselves.

Ornate box turtles primarily live in western prairies and grasslands.

Females, like this one, usually have eyes with more brown or yellow than the bright red eyes of some males.

REPRODUCTION Ornate box turtles mate from April through October. Mating occurs on land. Females dig nests in areas of open soil with abundant sunlight in which they lay 2–8 eggs (average = 5). The female may also dig a shallow pit in which she rests her plastron while digging a nest cavity that is about 5–6 inches deep. Females lay one or two clutches per year. Although ornate box turtles are active primarily during the day, females occasionally construct their nests at night.

PREDATORS AND DEFENSE The list of potential terrestrial predators of this species in the Southeast includes many carnivorous mammals and birds, such as raccoons, coyotes, skunks, crows, and ravens. The turtle's primary defense against predators is to withdraw into its shell and close the plastron up against the carapace. Ornate box turtles are vulnerable to sudden cold snaps, which can kill individuals that are not buried deep enough underground.

CONSERVATION ISSUES Loss of native prairies to agriculture has undoubtedly eliminated much of the species' original habitat. Large numbers are killed on the roads that fragment the remaining patches of suitable habitat. Ornate box turtles may have large home ranges and seek different areas for foraging, mating, nesting, and hibernation. As the remaining patches of suitable habitat shrink, the turtles are forced to cross unsuitable, human-altered habitats where encounters with dogs and raccoons are more likely. Fire suppression in grassland and prairie habitats favors the growth of forests, which are unsuitable habitat for these turtles. Ornate box turtles have long been collected as pets; some are even shipped overseas to Europe. Large-scale collection of these long-lived animals results in decline of native populations. Habitat management of oak-savanna woodlands and prairies on publicly owned land should incorporate the needs of ornate box turtles. The species is considered rare in Arkansas and is at risk of being extirpated from that state.

Eastern box turtles have perhaps the most variable color patterns of the four subspecies.

How do you identify a common box turtle?

DISTINGUISHING CHARACTERS
Can completely close up inside high-domed shell; pattern very variable; usually found on land

AVERAGE SIZE

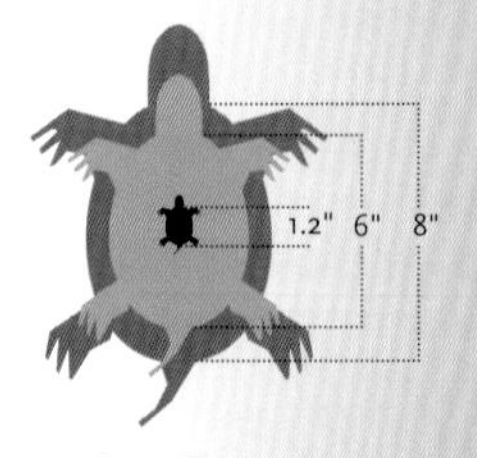

● ADULT—*T. c. major*
● ADULT—*T. c. carolina, T. c. triangius, T.c. bauri*
● HATCHLING

CARAPACE SHAPE

FRONT VIEW

SIDE VIEW

TOP VIEW

Common Box Turtle *Terrapene carolina*

DESCRIPTION Virtually no other turtle of the Southeast displays as much regional variation in color, size, shell shape, and habitat use as the common box turtle. The species is well known to anyone who spends time outdoors in the woods and fields of the Southeast. Common box turtles have a high-domed, frequently colorful carapace that often has the remnants of a keel down the center, a hinged plastron within which the head and limbs can be completely enclosed, and toes with no webbing. The carapace is usually brown or black, sometimes olive-brown, with markings of yellow and often orange or red. The plastron is also highly variable in color, ranging from yellowish to brown or black, sometimes with dark markings and sometimes without. Males in some regions have red eyes while females have brown eyes. Adult males have a pronounced depression in the center of the plastron.

VARIATION AND TAXONOMIC ISSUES Of the six subspecies of the common box turtle, four are found in the Southeast; all show extensive regional variability in color pattern. The eastern box turtle (*T. c. carolina*) has the greatest geographic range and the most variable shell coloration, usually a mix of colors including shades of yellow, orange, brown, and black. The shell pattern can be spots, dashes, or rhomboids, often with an intricate design on each scute. The eyes of most adult males are red and those of

Common Box Turtle

Terrapene carolina

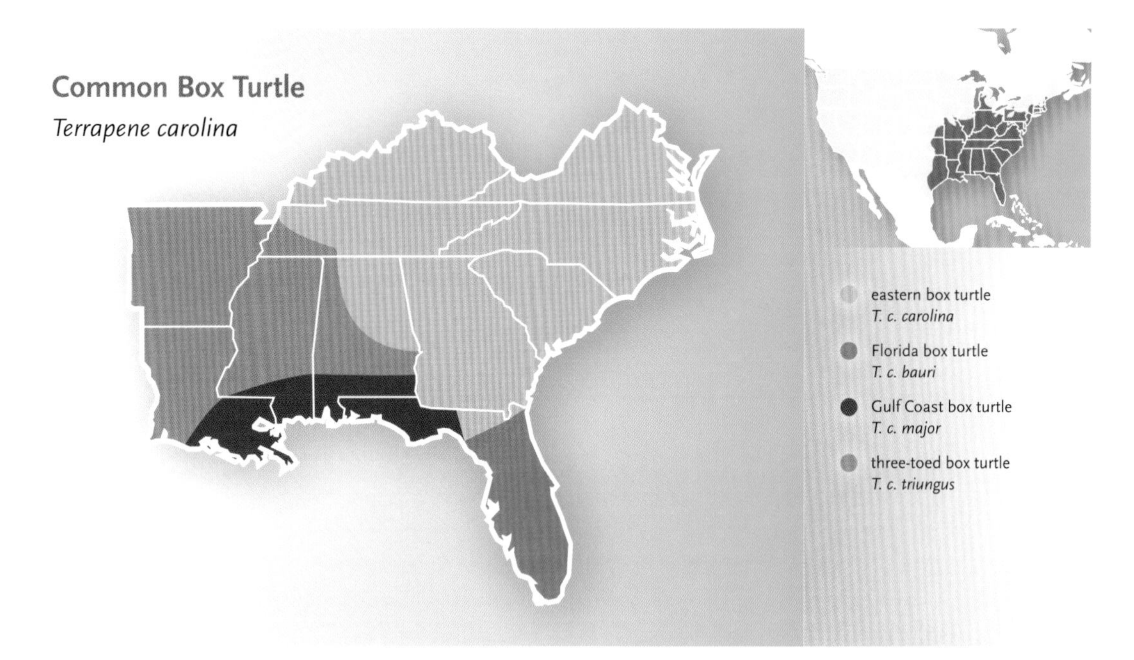

Eastern box turtle (left), Gulf Coast box turtle (top right), male three-toed box turtle (middle right), Florida box turtle (bottom right)

females are brown, although an occasional female has red eyes. The Gulf Coast box turtle (*T. c. major*) is the largest. The rear of its carapace has a distinctive flare, and males often have large white patches on the head. Eastern and Gulf Coast box turtles have four toes on each hind foot. Three-toed box turtles (*T. c. triunguis*) have a rather drab, horn-colored carapace, but the head has bright spots of red, orange, yellow, and occasionally even blue. The head of males is often almost solid red. The Florida box turtle (*T. c. bauri*) is the most distinctive of the four southeastern subspecies, having radiating yellow lines on each scute of the carapace that superficially resemble those of the ornate box turtle. Two yellow parallel stripes or dashes on the head and a yellow stripe down the center of the carapace are also typical of this subspecies. The eyes of the male Florida box turtle are brown. Florida and three-toed box turtles usually have three toes on the hind feet, but a few individuals have four. Intergradation occurs along the zones where the various subspecies come into contact, resulting in a mix of colors and patterns that make subspecies identification difficult.

WHAT DO THE HATCHLINGS LOOK LIKE? Hatchlings generally are brownish gray with only a single yellow dot within each carapace scute. Hatchlings of the Florida subspecies are often striking yellow and black. The most distinctive feature is a clearly noticeable keel or line of raised bumps down the center of the carapace. The shell is not as domed as it is in adults, and the hinge of the plastron is nonfunctional.

The hatchling's most distinctive feature is a clearly noticeable keel or line of raised bumps down the center of the carapace (top).

Hatchlings of the Florida subspecies (bottom) tend to be more brightly colored.

CONFUSING SPECIES Only one of the southeastern subspecies—the three-toed box turtle—overlaps the range of the ornate box turtle, which typically has four toes on the hind feet and yellow lines in each scute of the carapace, but lacks the distinct vertebral keel.

DISTRIBUTION AND HABITAT The subspecies of the common box turtle collectively occur throughout every southeastern state except a portion of southern Louisiana and occupy diverse habitats. They are found in mesic hardwood forests and fields in the Piedmont and mountains; in the Coastal Plain they inhabit sandy areas and palmetto thickets, wet meadows, pitcher plant bogs, and the borders of seasonal wetlands.

BEHAVIOR AND ACTIVITY Box turtles are active during the day throughout the warm seasons in the Southeast and year-round in the southernmost areas, preferring early morning in the summer. During severe drought or unusual heat waves, box turtles congregate in shallow pools, in moist areas of intermittent streams, or beneath overturned trees where moisture is retained. Individuals create shallow forms, or pits, beneath leaves and ground litter on the forest floor where they may spend the night during warm weather or become dormant during winter, often returning to the same location in subsequent years. As the weather turns colder, the turtle will insulate itself by digging a deeper form. Fire-scarred box turtles or their shells occasionally turn up in some habitats, particularly longleaf pine forests and pine flatwoods. Most individuals can probably survive the natural fires that occur frequently in such habitats, but if too much ground fuel has accumulated, the turtles can be killed because they do not bury themselves deep enough to escape the higher heat.

FOOD AND FEEDING Box turtles eat a broad array of plant material, including mushrooms, roots, flowers, seeds, berries, muscadine grapes, and a variety of grasses. They also eat any small animals they can capture, including earthworms, grubs, beetles, crayfish, frogs and toads, salamanders, snakes, and even birds.

REPRODUCTION Box turtles mate most commonly in spring, but some also mate during the summer and into the fall. A female box turtle can completely reject mating with a male by closing her shell, but the male goes through a series of courtship behaviors that include biting, scratching, and nudging the female's shell as well as displaying the colorful underside of his throat to induce her to accept him. Mating is usually on land, but the Gulf Coast subspecies sometimes mates in shallow water. Box turtles nest primarily in the evening but will also do so in the morning. Females nest from May through July, laying one to two clutches of about 5 eggs each. The eggs hatch in early September, and most hatchlings emerge from the nest in the fall.

Female box turtles can store sperm from a single mating for up to 4 years, and can produce a fertile clutch of eggs in each of those years without further mating. Hatchling box turtles are very secretive and are seldom seen before they are 2 or 3 years of age. Perhaps they frequent subterranean passageways and leaf litter where they are protected from most predators.

Did you know?

A female turtle can store sperm and fertilize her eggs for several years after a single mating.

Box turtles are omnivores and will eat any small animal they can capture.

Box turtles often enter damp or shallow aquatic habitats during dry spells.

This long-lived box turtle is vulnerable to landscape changes that fragment its home range and separate its favorite habitat patches.

Common box turtles are frequent victims on roads in the Southeast.

PREDATORS AND DEFENSE A variety of predators destroy the nests of box turtles and feed on the eggs; raccoons, skunks, foxes, snakes (scarlet snakes and kingsnakes), and crows are among them. Crows are also known predators on juvenile box turtles, as are black racers and copperheads. Other bird predators on the young include Mississippi kites, egrets, and barn owls. Large dogs and coyotes occasionally kill adults. Because common box turtles in the Southeast do not bury themselves deep beneath ground litter, unusual cold snaps with several days of subfreezing weather sometimes kill them or make them easy prey for mammalian predators.

CONSERVATION ISSUES Habitat fragmentation and loss is the largest source of mortality for box turtles. Box turtles are long-lived animals with established home ranges and favorite habitat patches (e.g., a blackberry thicket or small vernal pool) that they may visit each year at similar times. When roads fragment the box turtle's habitat, the associated mortality can eventually eliminate populations. Successful strategies for maintaining box turtles and other wildlife in a human-modified landscape will require passageways that allow the animals to move under or over roads. Collection for the pet trade—including shipping overseas—has had severe impacts on local populations. Individuals from populations exposed to pesticides may develop severe ear abscesses; the health consequences are unknown. Respiratory illnesses similar to those found in gopher tortoises have been identified in box turtles.

SEMIAQUATIC TURTLES

The heavily sculptured shell and bright orange skin make the wood turtle one of the most attractive turtles of the Southeast.

Wood Turtle *Clemmys insculpta*

DESCRIPTION Wood turtles have a rough brown to grayish brown carapace with prominent growth lines on each scute giving the shell a chiseled look. The hingeless plastron is yellow with black blotches on the outer edge of each scute. The head is mostly black above, while the skin on the throat, neck, and lower legs varies from yellowish orange to bright orange. Adult males are usually larger than females and have a concave plastron, a much thicker tail, and enlarged scales on the front legs.

VARIATION AND TAXONOMIC ISSUES No subspecies of the wood turtle have been named, and no variation occurs within the limited southeastern portion of its geographic range. Some turtle biologists place the wood turtle and its close relative the bog turtle in the genus *Glyptemys*.

WHAT DO THE HATCHLINGS LOOK LIKE? Hatchling wood turtles lack the orange color of the adults and are basically black and brown or tan, with noticeably long tails.

Young wood turtles are typically gray and lack the orange coloration of adults.

How do you identify a wood turtle?

DISTINGUISHING CHARACTERS
Carapace with sculpted wood appearance; orange wash on legs

AVERAGE SIZE

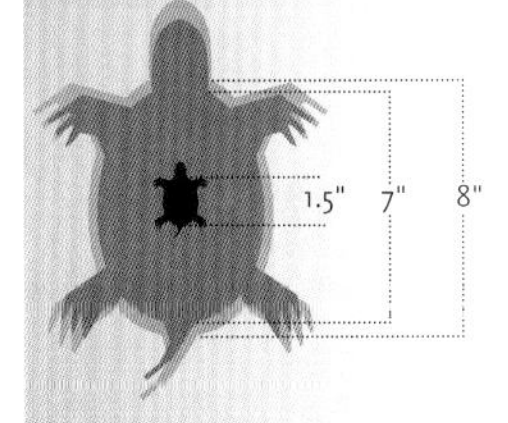

● FEMALE
● MALE
● HATCHLING

CARAPACE SHAPE

FRONT VIEW

SIDE VIEW

TOP VIEW

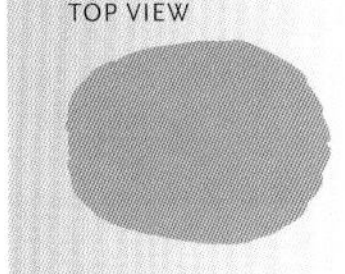

Wood turtles are terrestrial during summer months and forage in floodplain forests and fields.

CONFUSING SPECIES The combination of a dark head without stripes or prominent spots, a plastron with no hinge, and orange coloring on the neck and legs distinguishes wood turtles from all other North American turtles.

DISTRIBUTION AND HABITAT In the Southeast, the wood turtle is found only in Virginia, usually in landscapes with flowing, clear streams with sand or gravel and cobble bottoms; adjacent hardwood forests with diverse understory vegetation; and old fields with favorite summer foods such as strawberries and raspberries. Ponds, swamps, and other aquatic habitats with soft muck substrates are generally avoided.

BEHAVIOR AND ACTIVITY Wood turtles are often found in forests and fields during summer, where they behave much like common box turtles, although they rarely venture far from flowing water. During autumn, they return to the clear streams and spend the winter hibernating on the bottom of pools or beneath stream banks where tree roots provide cavelike refuges. Wood turtles tend to use the same foraging and hibernating sites year after

The wood turtle's plastron has no hinge.

Wood Turtle
Clemmys insculpta

year. They also show strong homing behavior and will try to return to their original location if displaced. Male wood turtles are territorial and are often aggressive toward other males.

FOOD AND FEEDING Wood turtles are omnivores that feed both on land and in the water. They eat mushrooms, flowers, grasses, and leaves—including poison ivy—on the forest floor. The same individual will visit a favorite blackberry, strawberry, or raspberry patch annually. Wood turtles eat many kinds of animals as well, including earthworms, grubs, other insects, snails, tadpoles, and dead fish. They sometimes exhibit an unusual behavior in which they thump the plastron on the ground and bring earthworms to the surface, where they are eaten.

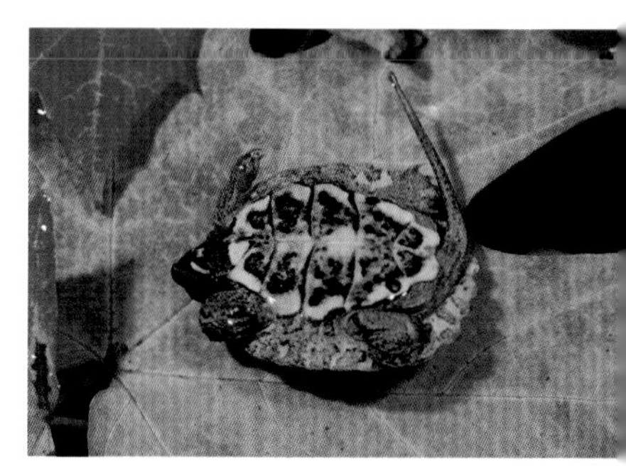

This hatchling shows characteristic plastral mottling and long tail (bottom), but lacks the chiseled carapace (top) that develops later in life.

REPRODUCTION Wood turtles take 10–20 years to reach maturity, and females may lay eggs for 50 years or more. Adults mate in their home streams in fall and spring before and after hibernation. Females lay a single clutch of 4–18 eggs (average = 10). Nesting females search for areas of open soil, including old gravel borrow pits and adjacent old

Wood turtles are most often found in or around clear rocky streams.

Orange skin on the neck and limbs is prominent on wood turtles.

fields or gravel bars that are above flood lines. They often nest in the evening during late May and June. Hatchlings emerge from the nest in late summer or autumn. They may wander some on land, but eventually they find their way to the edge of a stream, where they hide under objects and among streamside grasses.

PREDATORS AND DEFENSE Because they spend much of their time on land, wood turtles are exposed to a broad array of both terrestrial and aquatic predators. A high proportion of adults may be missing limbs or have injuries caused by carnivorous mammals, especially raccoons, which are also capable of killing juvenile and adult turtles. River otters and minks may kill wood turtles they find hibernating in streams.

CONSERVATION ISSUES Wood turtles are seriously threatened by habitat fragmentation. Because these turtles use streams, woodlands, and fields in their annual activity cycle, they are extremely vulnerable to automobile mortality, particularly in areas where roads parallel streams in valleys. Collectors can wipe out a wood turtle population in a few days by collecting the adults found along a stream. Adults foraging in agricultural fields are vulnerable to injury or death from heavy farming equipment.

A bog turtle walks across moist vegetation, beneath which it can readily conceal itself if threatened.

Bog Turtle *Clemmys muhlenbergii*

DESCRIPTION Adult bog turtles are among North America's smallest and most difficult to find turtles. The carapace is brown to rich mahogany in color, and the hingeless plastron is usually dark brown to black, with yellow patches that are often more prominent in younger individuals. Juveniles and young adults may have grooved concentric rings on the carapace scutes, but the carapace of older individuals may be worn nearly or completely smooth. A large yellow, orange, or coral red patch on both sides of the head is the identifying characteristic of the species.

VARIATION AND TAXONOMIC ISSUES The bog turtle has no subspecies, and no regional variation has been noted. Some turtle biologists place the bog turtle and wood turtle in the genus *Glyptemys*.

WHAT DO THE HATCHLINGS LOOK LIKE? Hatchling bog turtles look similar to the adults.

A hatchling displays a bright orange head patch.

How do you identify a bog turtle?

DISTINGUISHING CHARACTERS
Small; large yellow or orange blotches on sides of head

AVERAGE SIZE

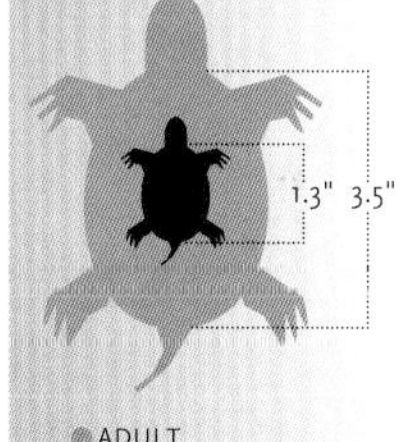

● ADULT
● HATCHLING

CARAPACE SHAPE

FRONT VIEW

SIDE VIEW

TOP VIEW

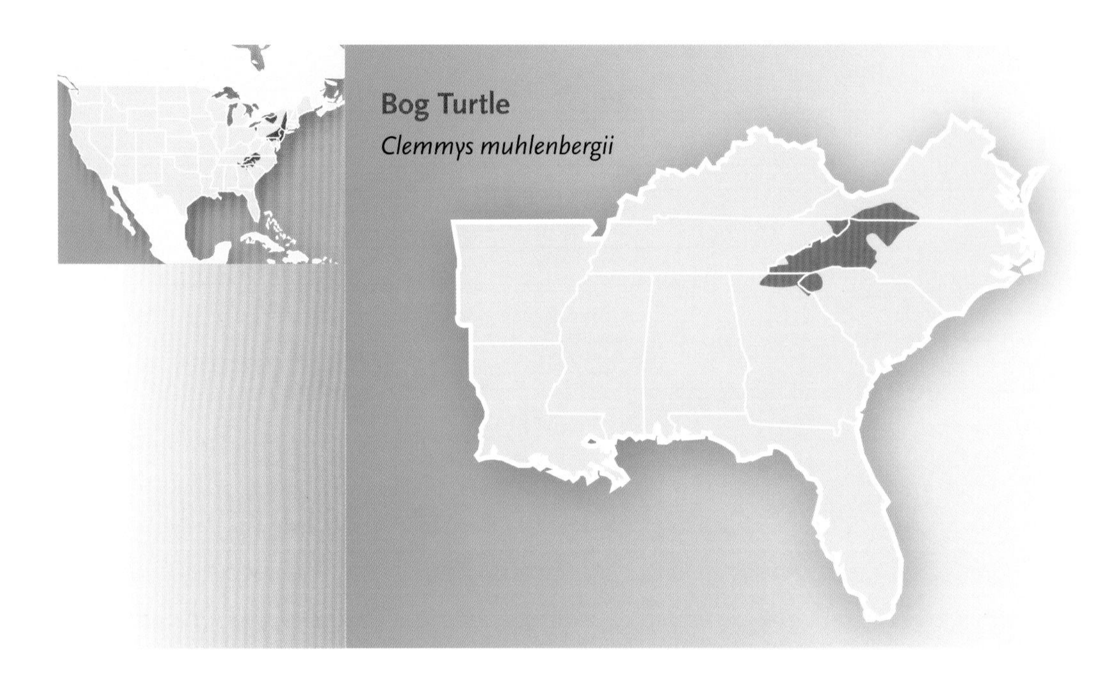

CONFUSING SPECIES The lack of a hinge on the plastron separates bog turtles from mud and musk turtles. The spotted turtle, which is probably the most likely species to be confused with the bog turtle, usually has yellow spots on the shell and several yellow spots on the head rather than only one. The carapace of adult bog turtles is typically higher domed than that of spotted turtles.

DISTRIBUTION AND HABITAT Bog turtles are known from only a few localities, usually in small numbers, mostly on the Blue Ridge Plateau in southwestern Virginia and in the Blue Ridge Mountains of the Carolinas, Tennessee, and Georgia. In North Carolina, several colonies are known from the Piedmont. Typical bog turtle habitat is open-canopy, wet meadows dominated by sedges, sphagnum moss, thick layers of organic muck, and flowing rivulets that are always associated with springs. Such habitats often exist along streams but not in isolated wetlands, which do not usually have bog turtle populations. Closed-canopy forest habitats do not allow adequate sunlight to reach the floor of the wetland for basking and egg incubation.

BEHAVIOR AND ACTIVITY Bog turtles are very secretive and spend much of their time concealed in mud. They often forage and travel under the cover of the wet meadow vegetation, using tunnels created by overhanging grasses and by small mammals such as meadow voles. Most bog turtles are discovered when they are seen basking in a shallow depression or

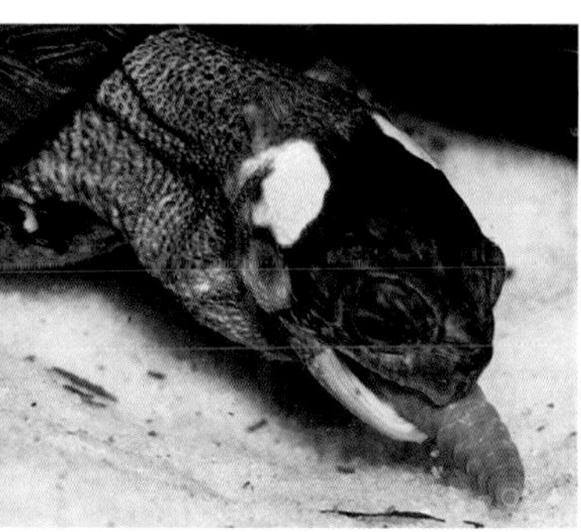

This bog turtle plastron (above) clearly exhibits growth annuli.

A bog turtle (left) consumes an earthworm.

atop a grass tussock. More frequently, they bask cryptically, hiding under vegetation and obtaining warmth without revealing their location. Although they may remain active during warmer parts of the year, bog turtles are most commonly observed in early spring when the vegetation is still low.

FOOD AND FEEDING Bog turtles are omnivores that capture live prey as well as scavenge, both on land and in the water. Among the plant materials consumed are seeds, berries, cattails, skunk cabbage, and duckweed. They will also eat baby birds that fall from nests, newborn meadow voles and mice, aquatic and terrestrial insects, crayfish, tadpoles, frogs, and salamanders.

Bog turtles' mahogany-colored shells and small size allow them to maneuver along muddy channels and conceal themselves beneath thick wetland vegetation.

REPRODUCTION During the spring mating season, males aggressively pursue females, even biting the female's legs, head, and neck to restrain her. Bog turtles sometimes lay their eggs in open terrestrial areas adjacent to wetlands that receive full sunlight, but most females nest in the top of sedge tussocks or in the thick sphagnum moss that lines the edges and floor of the most pristine wet meadow habitats. Each female lays a single clutch of 1–6 eggs (average = 3) in late April to early June. The eggs hatch in late August, and the hatchlings emerge soon afterward.

PREDATORS AND DEFENSE Bog turtles are often found with missing limbs that can be attributed to encounters with shrews, weasels, raccoons, foxes, skunks, or other mammalian predators. These little turtles often share their wet meadow homes with snapping turtles and may not survive a chance encounter. Because they are small and shy, and normally do not bite or scratch when picked up, they are defenseless against most large predators.

CONSERVATION ISSUES The principal conservation threat to bog turtles is the loss of the open, spring-fed, wet meadows they seem to require. Historically, such wet meadows appeared, disappeared, and reappeared on the landscape through a variety of natural processes. Ditching and drainage of wet meadows for farming and housing contributed to the decline of the species. Ironically, some human influences, such as cattle, goat, and sheep grazing, may have benefited the bog turtle by maintaining the open character of the wet meadows that would otherwise have become closed-canopy red maple swamps. Many bog turtle habitats can be restored and managed if private landowners and farmers, on whose land the majority of bog turtle colonies occur, are willing to work with nonprofit groups and state and federal agencies. Collection for the pet trade has also caused documented declines of bog turtles at several sites. Captive breeding programs could potentially fulfill the pet trade demands and also produce hatchlings to be reintroduced in sites where bog turtles previously occurred.

Females have yellow faces.

Spotted Turtle

Clemmys guttata

DESCRIPTION The spotted turtle is one of our most beautiful and gentle turtles. The solid black carapace is variously spotted with yellow dots, whose number can vary greatly, ranging from none to dozens. The yellow to orange plastron has a dark blotch on each scute and no hinge. The head is black and marked by yellow dots and often a larger dash mark. Some individuals have salmon-colored legs. Males have red eyes, gray faces, and concave plastrons; females have yellow eyes and face, and the plastron is flat.

VARIATION AND TAXONOMIC ISSUES No subspecies or geographic variation is apparent in the Southeast.

WHAT DO THE HATCHLINGS LOOK LIKE? Hatchlings resemble the adults except that each carapace scute usually has a single large yellow dot.

This juvenile still retains the hatchling pattern of one spot per scute.

How do you identify a spotted turtle?

DISTINGUISHING CHARACTERS
Black shell with yellow spots

AVERAGE SIZE

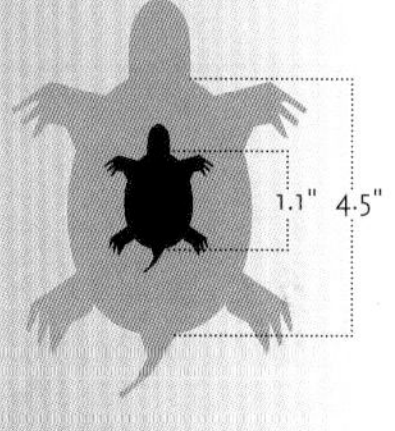

ADULT
HATCHLING

CARAPACE SHAPE

FRONT VIEW

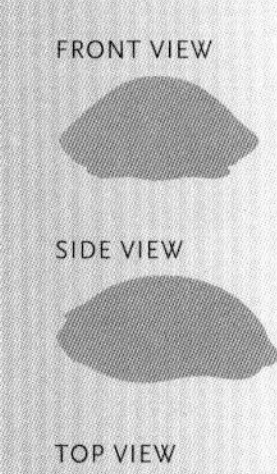

SIDE VIEW

TOP VIEW

Black and yellow blotches are characteristic of the plastron.

A female spotted turtle in a typical shallow-water habitat.

A male from South Carolina.

CONFUSING SPECIES The mud and musk turtles have hinged plastrons and more domed carapaces. Bog turtles have a yellow or orange blotch on each side of the head that is larger than the spots of the spotted turtles and lack spots on the carapace.

DISTRIBUTION AND HABITAT Spotted turtles are most abundant on the Coastal Plain and Piedmont; they are uncommon on the Florida peninsula. They are inhabitants of permanent wetlands, including wet meadows, grassy freshwater marshes, old beaver swamps, and the shallow edges of floodplain swamps, especially those consisting of old-growth cypress and tupelo swamp forest. They occasionally visit seasonal wetlands and may be encountered in small blackwater streams, which they use as corridors between wetlands. Classic habitat consists of shallow, clear or tannin-stained water interspersed with sedge tussocks and covered in duckweed. Spotted turtles avoid reservoirs, farm ponds, and other deep, open-water habitats that may be laden with sediments. The frequent association of spotted turtles with wetlands that have duckweed has led to speculation that the pattern of yellow spots serves as camouflage.

BEHAVIOR AND ACTIVITY Spotted turtles are noted for their seasonal activity in the Southeast, being most apparent in early spring. In cold areas they can often be found basking on sedge tussocks when snow remains on the ground and ice covers parts of the wetland. During summer they may continue to be active but are more difficult to find. Spotted turtles may

Did you know?

Bog turtles and spotted turtles may move less than a mile during their lifetimes, whereas some marine turtles may travel thousands of miles.

Spotted turtles occasionally have few or no spots.

aestivate in surrounding forests in some areas, especially during droughts or if they inhabit seasonal ponds that become dry. Both males and females will travel overland between aquatic habitats, including annual treks to hibernation sites.

FOOD AND FEEDING Spotted turtles are omnivorous. Plant food includes algae and wetland grasses. Small animal prey includes insects, snails, tadpoles, salamanders, and fish. They will also scavenge on dead fish or other carrion in the water.

REPRODUCTION Adults may congregate during the early spring mating season and engage in courtship that involves males pursuing females both in the water and onto adjacent land. The females may nest on dry land, but they often build nests in the tops of sedge tussocks, in woody debris, and in rotten logs within their shallow wetland habitats, and few people see them laying eggs. They usually nest one or two times in a given year, laying 1–8 eggs each time. The eggs hatch in mid-August, and some hatchlings overwinter in the nest.

Female spotted turtles sometimes bask or lay eggs on top of tussocks in small wetlands.

PREDATORS AND DEFENSE Because they seldom bite and have a relatively flat shell, spotted turtles are vulnerable to predators. Their primary defenses are concealment and secretive behavior, but it is not unusual to find spotted turtles with missing feet and limbs or with gnawed shells that are evidence of encounters with carnivorous mammals. In addition to raccoons, which are probably their most prevalent predator, other known predators are bald eagles and skunks.

CONSERVATION ISSUES A major conservation threat to spotted turtles is wetland loss and degradation. The small, shallow wetlands that harbor spotted turtles are often destroyed during construction of housing developments. Such habitats have also traditionally been drained for agricultural uses. Spotted turtles are favorites in the international pet trade, and collection of adult turtles could cause local extinctions. Captive breeding of spotted turtles for the pet trade is relatively easy, and if promoted, managed, and regulated would greatly reduce the pressure to remove spotted turtles from the wild.

Males typically have a gray face and red eyes.

A mottled head pattern and brown unmarked shell identify an adult eastern mud turtle.

How do you identify a common mud turtle?

DISTINGUISHING CHARACTERS
Can completely close up inside small, dark, rounded shell

AVERAGE SIZE

CARAPACE SHAPE

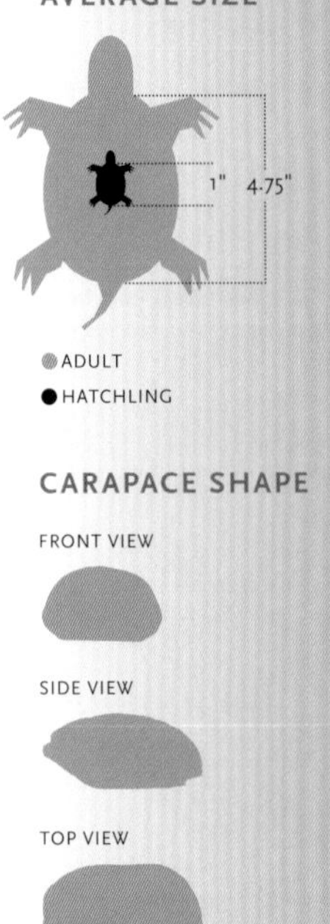

Common Mud Turtle *Kinosternon subrubrum*

DESCRIPTION A fist-sized adult common mud turtle characteristically has an olive-brown to black carapace, a light brown to yellowish plastron, and dark skin. The plastron is double hinged, which allows most individuals to close up tightly within the shell in the fashion of a box turtle. The tail of males is noticeably longer, and the notch at the rear of the plastron is considerably more pronounced to accommodate the thick tail. The male's tail has a curved, clawlike tip that he uses to help position himself during mating.

VARIATION AND TAXONOMIC ISSUES Adults of the three subspecies have similar shell colors, but other distinguishing features differentiate the geographic races. The head of the eastern mud turtle, *K. s. subrubrum*, and the Florida mud turtle, *K. s. steindachneri*, usually has no stripes but does have yellow mottling, whereas

The head of the eastern mud turtle (top) is typically mottled; the Mississippi subspecies (bottom) has yellowish stripes.

Common Mud Turtle

Kinosternon subrubrum

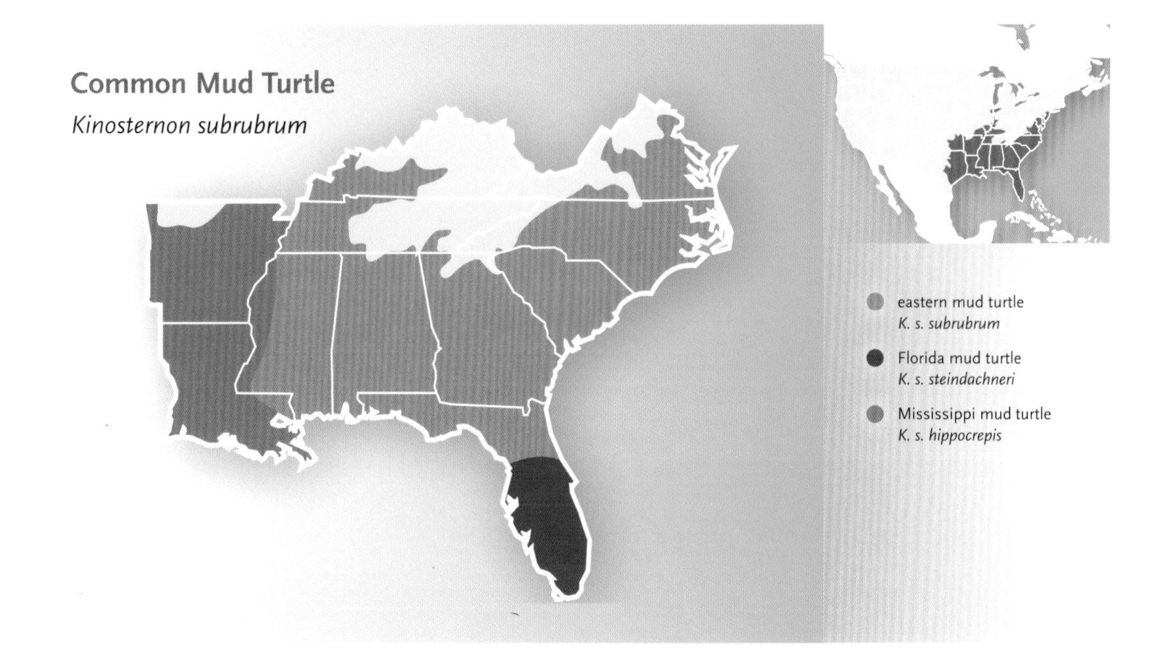

the Mississippi subspecies, *K. s. hippocrepis,* has a pair of yellowish stripes along each side of the dark brown head. The Florida mud turtle has a proportionately smaller plastron than the other two subspecies, and the males reach larger sizes and have larger heads than females.

A lack of head patterning and overall larger body size help to characterize this Florida subspecies of the common mud turtle (left top).

The Mississippi mud turtle has a pair of yellowish stripes along each side of the dark brown head (left bottom).

WHAT DO THE HATCHLINGS LOOK LIKE? Baby mud turtles resemble small walnuts. They are typically black above and have a red, yellow, or orange plastron with scattered areas of black.

CONFUSING SPECIES The striped mud turtle, which co-occurs with eastern and Florida mud turtles, has yellow head stripes that may be conspicuous only between the eye and nostril. Some eastern mud turtles have head stripes, but they are broken or jagged and do not occur between the eye and nostril. Mississippi mud turtles also have head stripes, but their geographic range does not overlap with that of the striped mud turtle. All common mud turtles have a two-

Common mud turtle hatchlings are quite small.

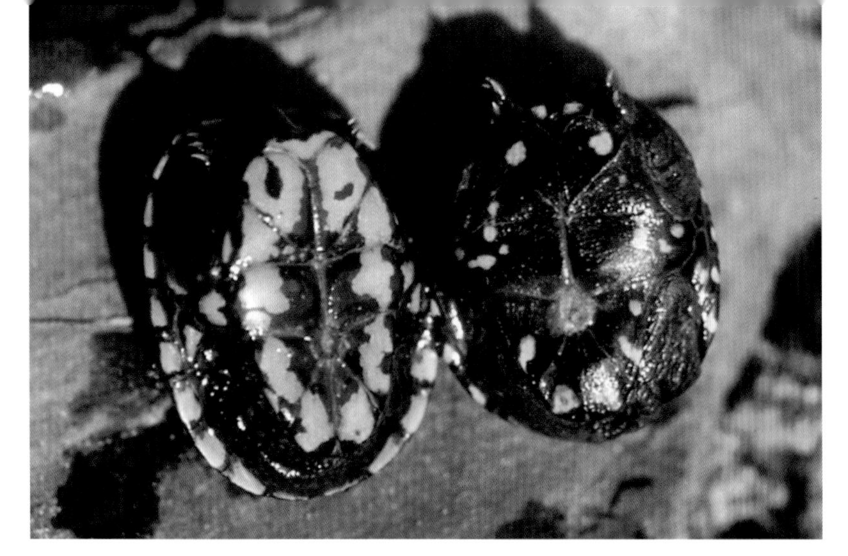

The plastrons of hatchlings of an eastern mud turtle (orange and black, *left*) and striped mud turtle (black with small yellow spots, *right*) illustrate some differences between the two species.

hinged plastron that closes tightly against the carapace. Musk turtles have one nonfunctional hinge and a much smaller plastron in relation to the size of the carapace.

Albino mud turtles, although rare, can occur in the same clutch as normal hatchlings.

DISTRIBUTION AND HABITAT Common mud turtles are found in every southeastern state in or around almost any aquatic habitat. They are most common in still bodies of heavily vegetated shallow water. Mud turtles in streams, rivers, and reservoirs are most likely to be in shallow backwater areas. The species is found on numerous barrier islands of the Atlantic and Gulf coasts, suggesting a tolerance to brackish water, although these turtles are not permanent residents of brackish tidal creeks.

BEHAVIOR AND ACTIVITY Like most other southeastern turtles, common mud turtles are active during warm periods of the year, with those in coastal and southern localities having longer periods of activity than northern and inland populations. Thus, the Florida subspecies may be active year-round, while the eastern mud turtle hibernates from November to March. Common mud turtles in many populations spend as much as half of each year on land adjacent to their aquatic habitat, generally buried beneath the surface, where they hibernate during cold weather and aestivate during dry periods. The Florida subspecies appears to be more aquatic than the other two and is found less frequently on land. Although considered semiaquatic, mud turtles are most commonly seen when they are traveling overland, often between a terrestrial hideaway and the aquatic habitat. They seldom bask on logs or on the shore.

FOOD AND FEEDING Mud turtles eat a variety of small invertebrates, seeds, algae, and aquatic vegetation. They also scavenge on dead fish and other carrion, and will prey on small frogs and salamanders if the opportunity presents itself.

REPRODUCTION The reproductive season begins with mating in early spring followed by nesting during spring and early summer. The reproductive season is more extensive in warmer localities and may occur year-round in parts of Florida. Mating has been observed both in the water and on land. Females lay from one to three clutches of 1–6 eggs each (average = 2–4), often during or immediately after rain. They dig a nest in sandy or loamy soil, deposit the eggs, and cover them with soil and sometimes dead vegetation. Eggs are occasionally laid beneath rotting logs or boards. Females may spend several days on land after nesting—perhaps to rest or to wait for favorable weather conditions to make the journey back to water. The eggs may incubate for more than 3 months, much longer than those of most U.S. turtles, and the hatchlings remain in the underground nest through the fall and into winter, emerging during a warm spell in late winter or early spring.

PREDATORS AND DEFENSE Despite their small size, adult mud turtles have few natural enemies and can live for 30 years in the wild, and perhaps much longer. Mid-sized carnivorous mammals such as raccoons, skunks, and foxes, and kingsnakes and subterranean predators such as shrews, will eat the eggs when they encounter a nest. Fire ants have been observed killing adults. Because of their small size during their first year, young mud turtles are vulnerable to a variety of wetland predators, including water snakes, cottonmouths, and perhaps large wading birds. Mud turtles characteristically try to bite when picked up. They can deliver a painful nip and may not let go immediately, but their initial response to humans or predators encountered on land is to close up inside the protective shell.

CONSERVATION ISSUES Highway mortality is a primary threat to mud turtles throughout their range because of their high propensity for overland movement and slow speed in crossing roads. Few mud turtles manage to cross a busy highway successfully. Wetland laws and regulations should include the upland habitat surrounding aquatic areas because these peripheral terrestrial areas are critical for much of a mud turtle's life cycle.

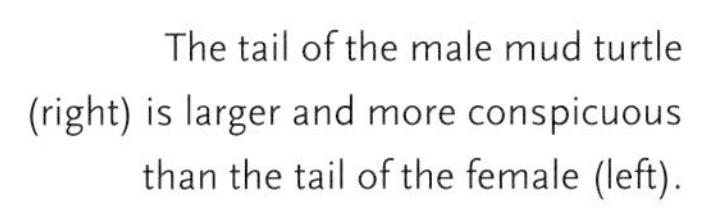

The tail of the male mud turtle (right) is larger and more conspicuous than the tail of the female (left).

The top yellow head stripe goes over and across the nose, and the carapacial stripes are visible in Florida specimens. The stripes may become less conspicuous in older individuals

How do you identify a striped mud turtle?

DISTINGUISHING CHARACTERS
Can completely withdraw into shell; stripe between eye and nostril; in Florida, three yellow stripes along length of carapace

AVERAGE SIZE

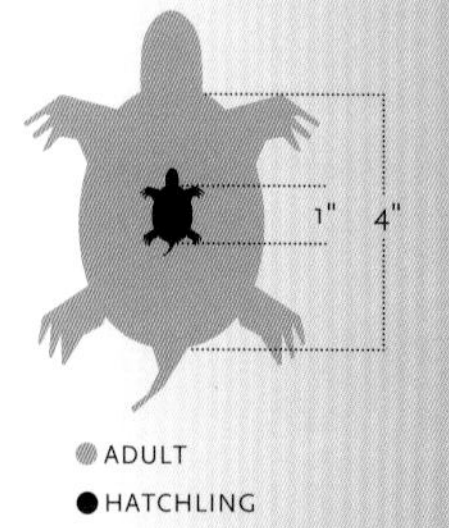

● ADULT
● HATCHLING

CARAPACE SHAPE

FRONT VIEW

SIDE VIEW

TOP VIEW

Striped Mud Turtle *Kinosternon baurii*

DESCRIPTION Striped mud turtles are oval, with a smooth, dark carapace that is brown or black; dark skin; and a light brown to yellowish plastron with a double hinge. In most of Florida the carapace has the three longitudinal yellow stripes (down the center and on each side) for which the species is named. Outside Florida, the yellow stripes are faint or completely absent, even on young specimens. Two yellow head stripes are visible on each side of the head in most specimens throughout the range, with the top stripe on each side continuing past the eye to the nostril. Females are slightly larger than males. Adult males have longer tails than females, and

A yearling displays its three carapace stripes.

Hatchling plastrons are often solid black, but occasionally they have small yellow spots (see page 78).

Striped Mud Turtle

Kinosternon baurii

Striped mud turtles from Georgia north to Virginia (left) do not have stripes on their carapace, although they retain the head stripes.

The long abdominal scute (right) on the striped mud turtle helps to distinguish it from common mud turtles.

the tail ends in a curved, clawlike tip that presumably helps the male hold on to the female during courtship and mating.

VARIATION AND TAXONOMIC ISSUES No subspecies have been described, but the color pattern varies geographically; the distinct three yellow shell stripes characteristic of most Florida specimens are obscure or even absent on specimens from elsewhere. A population in the Florida Keys lacks both shell stripes and head stripes.

WHAT DO THE HATCHLINGS LOOK LIKE? Baby striped mud turtles are black above and predominantly black below; the plastron occasionally has small spots or patches of yellow. The plastron has no hinges at birth, but these usually develop by 3 months of age.

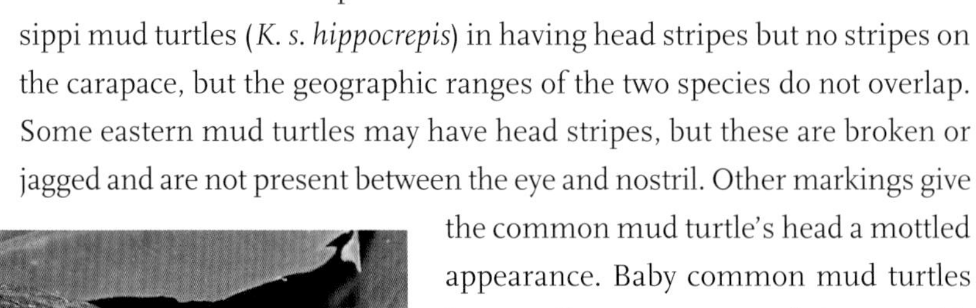

CONFUSING SPECIES Striped mud turtles north of Florida resemble Mississippi mud turtles (*K. s. hippocrepis*) in having head stripes but no stripes on the carapace, but the geographic ranges of the two species do not overlap. Some eastern mud turtles may have head stripes, but these are broken or jagged and are not present between the eye and nostril. Other markings give the common mud turtle's head a mottled appearance. Baby common mud turtles have a yellow, orange, or red plastron with black blotches.

Striped mud turtles from the Florida Keys lack head and carapace stripes.

DISTRIBUTION AND HABITAT Striped mud turtles are found in the Atlantic Coastal Plain from the Florida Keys to Virginia, in association with isolated seasonal wetlands and floodplain swamps that may be heavily vegetated. Individuals are found most commonly in and around backwater swamps and sloughs of rivers, frequently flooded ditches, and small sinkhole lakes.

Did you know?

Before hatching, the eggs of at least two species of turtles in the Southeast, the chicken turtle and the striped mud turtle, go through periods in which development completely stops for several weeks. They are also the only species whose eggs pass through the winter unhatched.

BEHAVIOR AND ACTIVITY Striped mud turtles are active throughout most of the year in central and southern Florida; the period of activity decreases during winter in northern and more inland localities. They spend major portions of the year on land, burrowed beneath soil and ground litter. As with the common mud turtle, striped mud turtles are most likely to be encountered as they travel overland to or from nesting or hibernation sites, or to or from aestivation sites in response to fluctuating water levels in their aquatic habitat. They are seldom seen basking out of the water.

FOOD AND FEEDING Striped mud turtles eat a variety of animal and plant material, including algae, seeds, and leaves of aquatic plants, as well as insects and mollusks. They presumably scavenge on dead fish and other animals as well.

REPRODUCTION The reproductive pattern of striped mud turtles is atypical in that females lay their eggs in fall and early spring or during warm spells in winter, although they may nest year-round in southern Florida. Females may also retain eggs for up to several months over the winter if air temperatures are too cold to allow them to nest. When seasonal wetlands dry in late summer, female striped mud turtles bury themselves on land

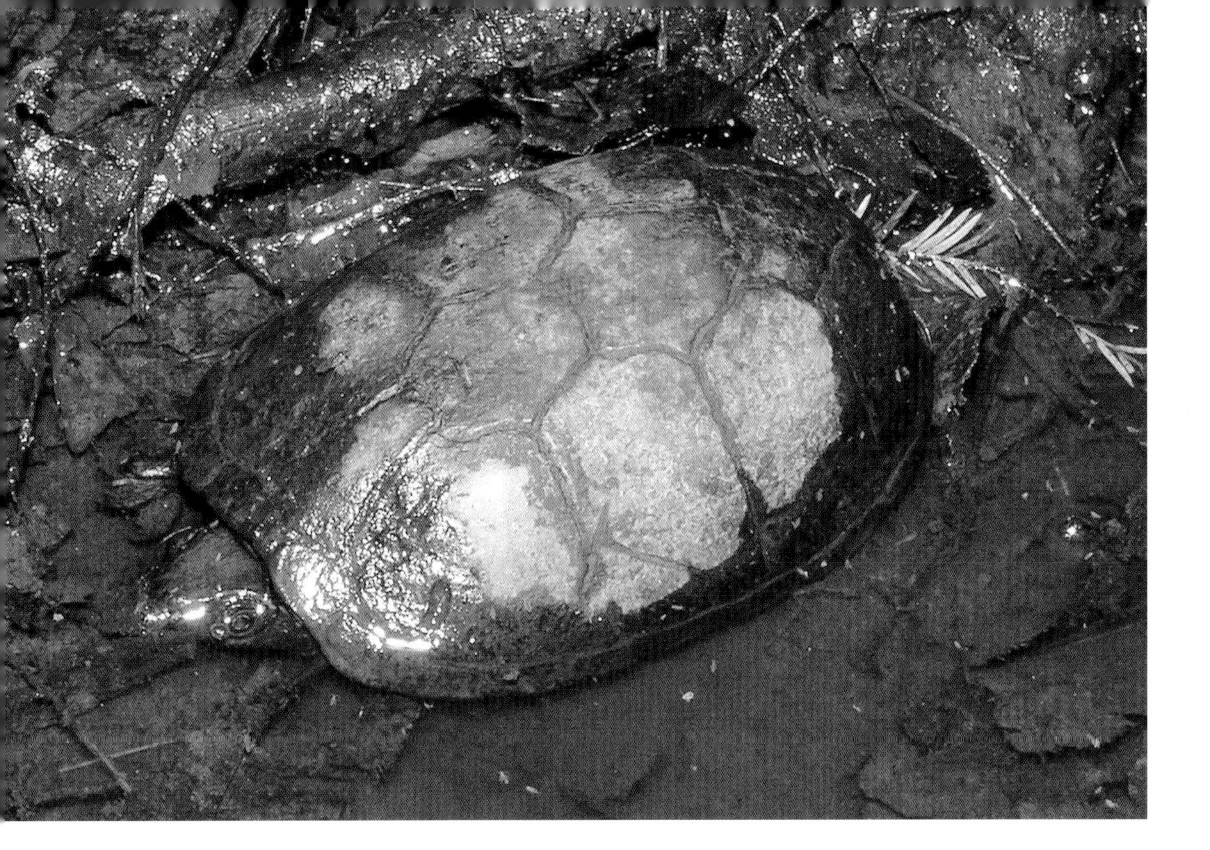

while retaining their eggs. The incubation period for eggs in the nest is usually at least 3 months; however, eggs laid in the fall may not begin developing until the following spring, so the time the eggs are in the nest can be longer. Clutches may contain 1–7 eggs (average = 2–4).

All mud turtles thrive in mucky habitats (top).

A prominently striped individual (bottom) typical of those from central Florida.

PREDATORS AND DEFENSE Aside from nest predation by mammals and kingsnakes, striped mud turtles in Florida are eaten by alligators and snail kites. Presumably, juvenile striped mud turtles are vulnerable to a wide variety of other aquatic and terrestrial predators because of their small size.

Striped mud turtles typically do not try to bite even when given an opportunity. When an individual is encountered on land, its first response is to close up inside the protective shell and to remain so when picked up.

CONSERVATION ISSUES Striped mud turtles are vulnerable to road mortality because of their frequent overland travel. As with common mud turtles, which also spend a considerable proportion of their life on land, the terrestrial habitat peripheral to aquatic areas is of key importance to this species and should be protected as part of any wetland where it lives. The population on the Florida Keys has been considered at special risk of extirpation from loss of habitat.

Orange edges on the carapace are most characteristic of chicken turtles from the Florida peninsula. All chicken turtles have a broad yellow stripe on the front legs.

How do you identify a chicken turtle?

DISTINGUISHING CHARACTERS
Long neck; shell pear-shaped from above; netlike pattern on carapace

AVERAGE SIZE

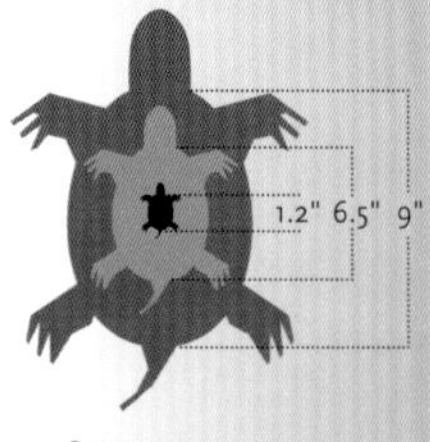

● FEMALE
● MALE
● HATCHLING

CARAPACE SHAPE

FRONT VIEW

SIDE VIEW

TOP VIEW

Chicken Turtle

Deirochelys reticularia

DESCRIPTION The chicken turtle is a medium-sized, semiaquatic, freshwater turtle that is unusual among other hard-shelled turtles of the Southeast in having a long neck and pear-shaped shell. Bright yellow stripes are visible on the head, neck, and legs, and a reticulate, or netlike, pattern of yellow lines is conspicuous on the carapace. The carapace is generally dark brown to black, although some shells are light brown or olive. The plastron is plain yellow or may have black marking in the seams. Black bars or an occasional spot may be present on the bridge between the plastron and carapace. The rear marginal scutes of the shell are smooth edged, not serrated. One usually broad yellow stripe is present on each front limb. Females are generally larger than males. The name "chicken turtle" may refer to the long, chickenlike neck but more likely comes from the first account printed about the species in the 1800s, which stated that "the turtle is more palatable than the cooter, and in fact tastes like chicken."

The wide flattened head of a female chicken turtle is suited for consuming crayfish, a favorite food.

Chicken Turtle

Deirochelys reticularia

The head and leg striping on western chicken turtles is more cream-colored than the yellow of the eastern subspecies.

A reticulate pattern on the carapace is visible on this eastern chicken turtle.

Hatchling chicken turtles resemble the adults.

VARIATION AND TAXONOMIC ISSUES Three subspecies have been described, and all occur within the Southeast. The eastern chicken turtle, *D. r. reticularia*, has a plain yellow plastron and a brown to black carapace with reticulate yellow lines. The Florida chicken turtle, *D. r. chrysea*, the largest of the three subspecies, is similar but with bolder yellow or even orange patterning on the carapace, especially the marginal scutes. The western chicken turtle, *D. r. miaria*, has black markings within the seams on the plastron, and the striping on the head is often pale yellow or cream colored.

WHAT DO THE HATCHLINGS LOOK LIKE? Hatchlings are patterned very much like the adults.

CONFUSING SPECIES Where they co-occur in the east, yellow-bellied sliders have a broad yellow mark behind the eye, and the rear of the carapace is serrated. Painted turtles have red stripes on the legs. Adult river cooters and pond cooters are larger than chicken turtles, and their carapaces often flare outward in the rear. The chicken turtle is the only species with all of the following characteristics: vertical yellow lines on the rump, a broad yellow bar on each forelimb, and a smooth carapace edge in the back.

DISTRIBUTION AND HABITAT Chicken turtles occur throughout the Southeast from southeastern Virginia and the Outer Banks of North Carolina through the Deep South and up the Mississippi Valley through eastern Arkansas. They are confined to the

Chicken turtles will leave the aquatic habitat to aestivate on land part of the year.

Atlantic and Gulf coastal plains and the Florida peninsula. Seasonal isolated wetlands, such as Carolina bays and sinkhole ponds, are the preferred habitat. Chicken turtles are not found in water with suspended sediments or clay but instead prefer the black or tea-colored water typical of Coastal Plain wetlands. They sometimes occur in floodplain swamp forests and have been reported from borrow pits and ponds. They do not occur in rivers or streams. Ponds with large predatory fish are generally avoided, as are water bodies with large alligators.

Black bars may be present on the bridge between a chicken turtle's plastron and carapace.

BEHAVIOR AND ACTIVITY Although widely regarded as an aquatic turtle, the chicken turtle spends many months of each year on land. Seasonal wetlands are often dry for a portion of each year, notably from late summer through winter. During these times, chicken turtles walk out of their wetland and into the surrounding forest, and dig themselves in under pine straw or hardwood leaves and moist—but not saturated—soil. Here they remain without feeding, inactive but alert for the rains that will refill the wetland.

FOOD AND FEEDING Chicken turtles are active predators that may rely more on sight than smell to find food. The juveniles and males eat dragonfly and damselfly larvae, and the large females actively hunt crayfish. Other aquatic insects, tadpoles, and salamander larvae are likely taken as well.

REPRODUCTION Chicken turtles mature quickly relative to other turtles, with males reaching maturity in as few as 2 years. Females first lay eggs when they are 5–6 years old. Nesting begins in August or September. Most females lay two clutches of 5–13 eggs

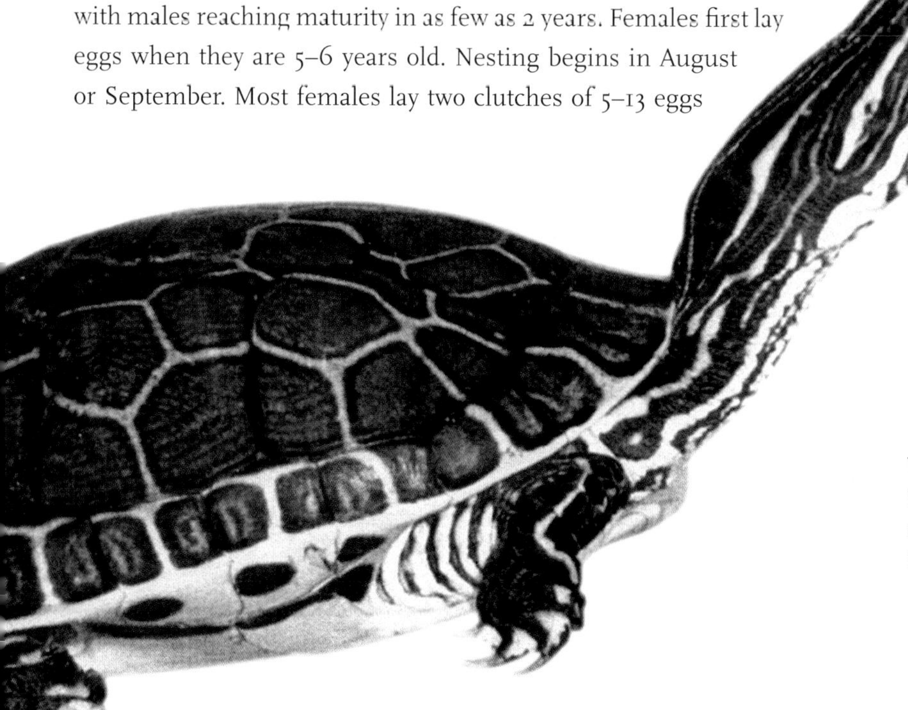

The chicken turtle uses its extremely long neck to capture crayfish—one of its favorite prey.

and often have to hold the second clutch into the winter, sometimes digging a nest on a warm winter day, and sometimes waiting until March. Regardless of when they are laid, the eggs do not start to develop until the ground warms up late the following spring. The eggs hatch in late summer. Hatchlings may wait in the nest chamber or may emerge and bury themselves in the surrounding forest until the following spring when the wetland is again full of water.

PREDATORS AND DEFENSE Raccoons are probably the most damaging predators, digging up nests and killing nesting females. Invasive fire ants will kill hatchlings, negating the chicken turtle's strategy of remaining in the nest until spring. Wading birds such as herons likely consume some hatchlings. Predators of adults also include otters and alligators. Chicken turtle shells are thinner than those of co-occurring species like yellow-bellied sliders, and they cannot survive in habitats occupied by large or even moderate-sized alligators.

Adult male chicken turtles (right) are noticeably smaller than adult females (left) but have larger tails.

CONSERVATION ISSUES The loss of seasonal wetland habitats throughout the Southeast has undoubtedly reduced the number of chicken turtle populations. Chicken turtles also require upland forested habitats around their seasonal wetlands where they can take refuge when the wetlands are dry. Wetland regulations that provide only narrow buffers to protect wetland water quality are insufficient to protect the fauna of seasonal wetlands, including chicken turtles.

This southern painted turtle has the red vertebral stripe that is characteristic of the subspecies and the long front claws and thick tail that are typical of adult males.

Painted Turtle *Chrysemys picta*

DESCRIPTION Painted turtles are among the most colorful turtles in the Southeast. The completely smooth shell is black, brown, or olive, often with a hatched pattern of yellow bars that run both horizontally and vertically on the carapace. The head has numerous bright yellow lines and occasionally blotches, and the black limbs have brightly colored red stripes. In addition to being smaller than females, adult males have conspicuously elongated front claws.

VARIATION AND TAXONOMIC ISSUES Three of the four recognized subspecies occur in the Southeast, but they are generally distinguishable by their color patterns and shell features. The eastern painted turtle, *C. p. picta*, has enlarged yellow spots behind each eye, and the plastron is usually immaculate yellow. Yellow striping on the head may break into one or more spots behind each eye or remain as continuous stripes in the midland painted turtle, *C. p. marginata*, or become jagged in the southern painted turtle, *C. p. dorsalis*. The center of the plastron of the midland subspecies has a dark blotch of variable size, and the southern form usually has an unmarked plastron. A typical southern painted turtle can be readily distinguished from all of the other subspecies by the bright red, or occasionally yellow, stripe that runs along the spine of the carapace from head to tail.

How do you identify a painted turtle?

DISTINGUISHING CHARACTERS
Black legs with red stripes

AVERAGE SIZE

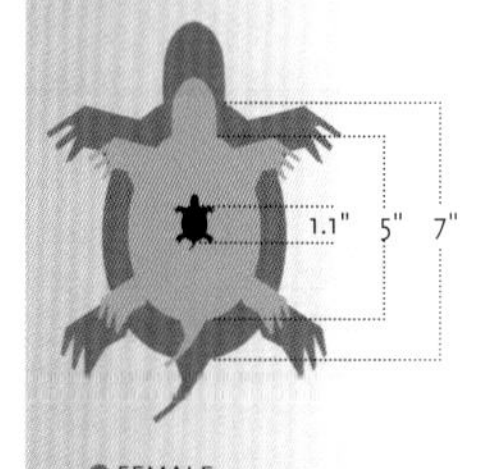

● FEMALE
● MALE
● HATCHLING

CARAPACE SHAPE

FRONT VIEW

SIDE VIEW

TOP VIEW

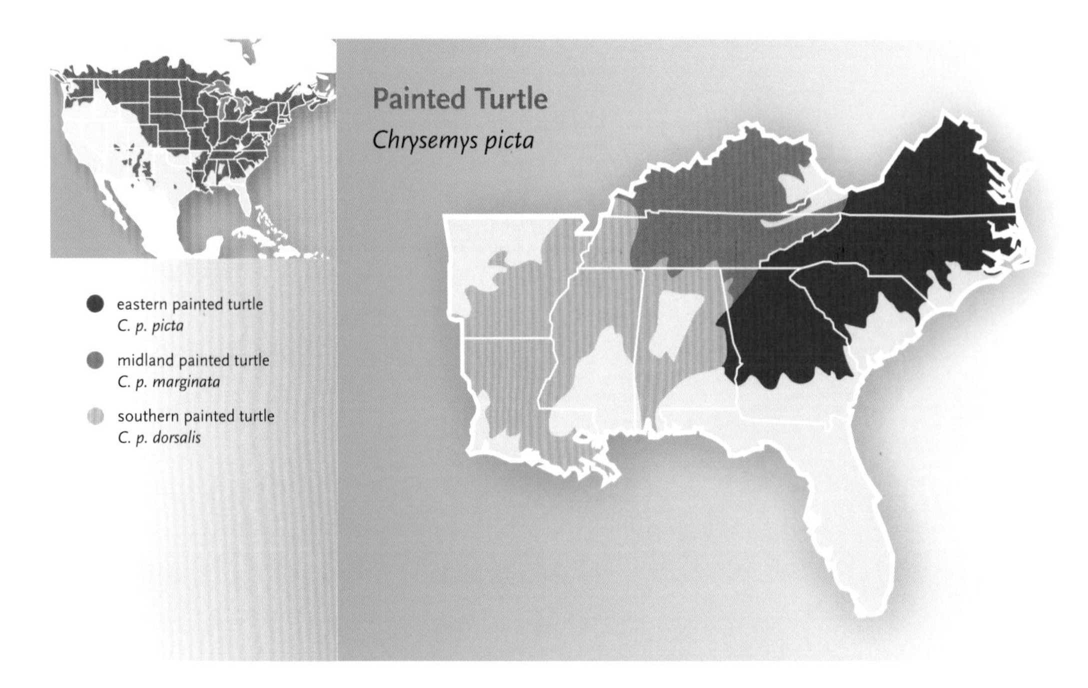

Some turtle biologists consider the southern painted turtle to be a separate species, *C. dorsalis.*

WHAT DO THE HATCHLINGS LOOK LIKE? Hatchlings are tiny replicas of the adults.

CONFUSING SPECIES The red-lined legs and the smooth, unserrated shell and margins set painted turtles apart from all other North American species.

DISTRIBUTION AND HABITAT Painted turtles are found in the mountains, Piedmont, and Atlantic and Gulf coastal plains, occurring in parts or all of every southeastern state except Florida. They can be found in a wide range of aquatic habitats, including farm ponds, lakes, and slow-moving rivers and adjacent swamps and oxbow lakes. They are frequently associated with beaver ponds and shallow marshes and occasionally with seasonal wetlands.

BEHAVIOR AND ACTIVITY Painted turtles are among the most commonly observed turtles because they bask throughout the year, even during winter on sunny days, and travel overland between aquatic sites during all of the warmer months. They are

Hatchling southern (top) and eastern (bottom) painted turtles are tiny replicas of the adults.

A midland painted turtle, one of the three subspecies found in the Southeast.

Two yellow spots on the head identify an eastern painted turtle.

frequently the most abundant species in recreational lakes and ponds and are often seen as they sit atop floating vegetation.

FOOD AND FEEDING Painted turtles will eat virtually anything. Their diet includes algae, duckweed, seeds, and vegetation as well as insects, worms, crayfish, snails, tadpoles and frogs, and carrion, especially dead fish.

REPRODUCTION Painted turtles mate in early spring and possibly again in the fall. Courtship starts with the male vibrating his long foreclaws in front of the female prior to actual mating. During the mating season, males frequently move overland between bodies of water in search of mates. During nesting, which takes place in May and June, females sometimes travel long distances to find open, sunny areas suitable for laying eggs. Nesting occurs during the day, and nesting females usually return to water before dark, possibly to avoid raccoons. One or two clutches of 1–8 eggs is typical. Although the eggs hatch by early autumn, hatchlings often remain in the nest chamber until the following spring.

Painted turtles from the New River in Virginia show characteristics of midland (left) and eastern (right) subspecies.

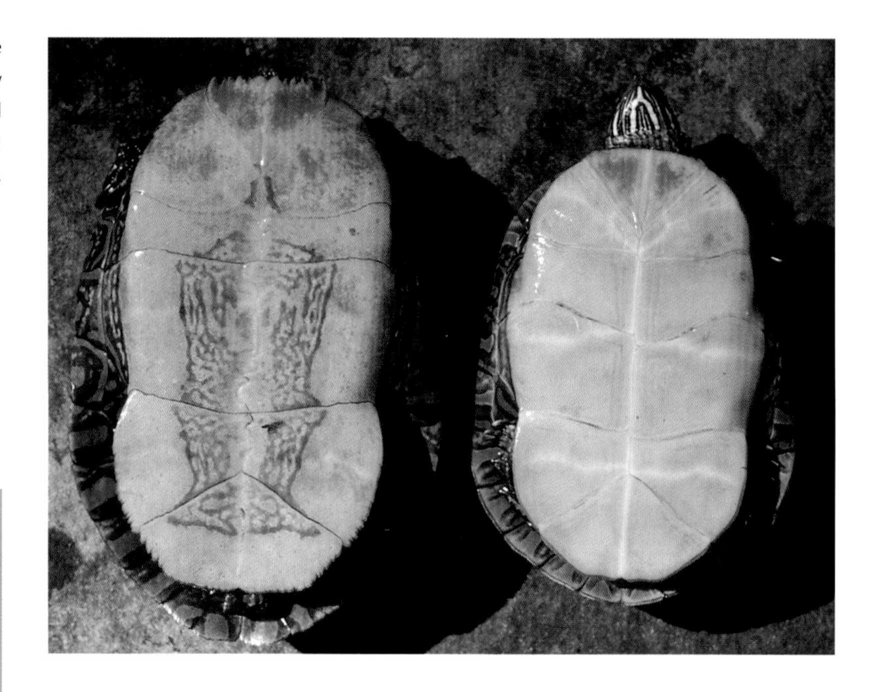

Did you know?

Even though turtle eggs of most species hatch at the end of summer, the hatchlings of some species may remain underground in the nest until the following spring.

PREDATORS AND DEFENSE Raccoons are the primary predators of painted turtle eggs, juveniles, and adults. Smaller painted turtles also fall prey to river otters, minks, herons, snapping turtles, wading birds, catfish, and watersnakes. Small painted turtles have been eaten by bullfrogs and killed by large predatory water bugs.

CONSERVATION ISSUES Painted turtles appear to be relatively secure from a conservation perspective. Their adaptability to man-made aquatic environments, including golf course ponds, farm ponds, urban lakes, and reservoirs, has allowed them to survive in altered landscapes. Ironically, their ability to inhabit areas populated by humans leads to special problems such as highway deaths, injuries from motorboat propellers, and predation of females and eggs by such "suburban predators" as raccoons and dogs.

The yellow-bellied slider is the eastern subspecies.

Slider Turtle

Trachemys scripta

DESCRIPTION The slider turtle is probably the most common and frequently observed turtle in the Southeast. This medium to large turtle has a relatively high-domed shell with a moderate keel down the center. The shell color is variable, ranging from greenish to brown or black or dark gray, often with broad yellowish bars, and the posterior edges of the carapace are usually distinctly jagged. The plastron is pale to deep yellow and usually has a dark spot on each of the front pair of scutes and sometimes on other scutes. The yellow marginal scutes also have solid dark spots on the underside. The dark head and legs have myriad yellow lines, and the line behind each eye enlarges to form a characteristic blotch that ranges from shades of pure yellow to pure red, depending on the geographic location. Adult males, which are smaller than adult females, have long foreclaws that are used in courtship. Old males become melanistic, with the colors on the head and legs fading and the skin appearing almost black. The carapace and the seams along the plastron become smoky gray or black.

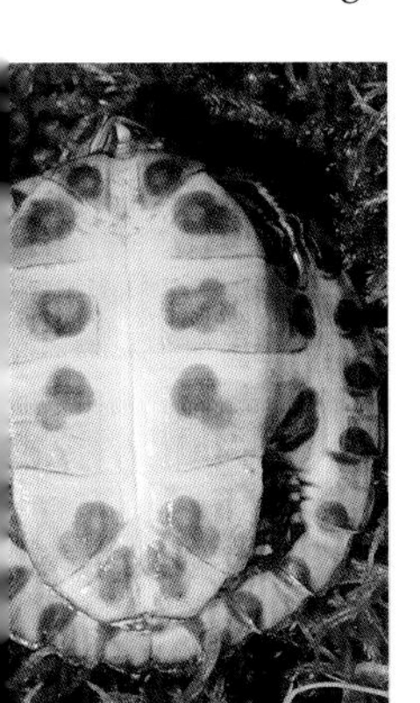

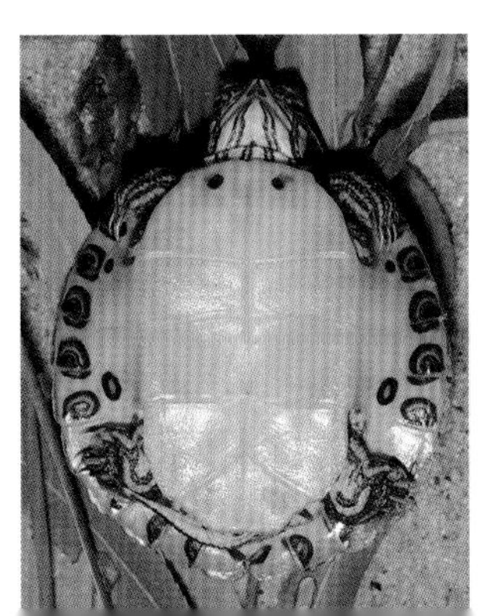

Red-eared sliders (left) have black smudges on most plastral scutes. Yellow-bellied sliders (right) have two small black spots on the anterior plastron.

How do you identify a slider turtle?

DISTINGUISHING CHARACTERS
Yellow or red blotch on side of head; yellow plastron with at least two black spots

AVERAGE SIZE

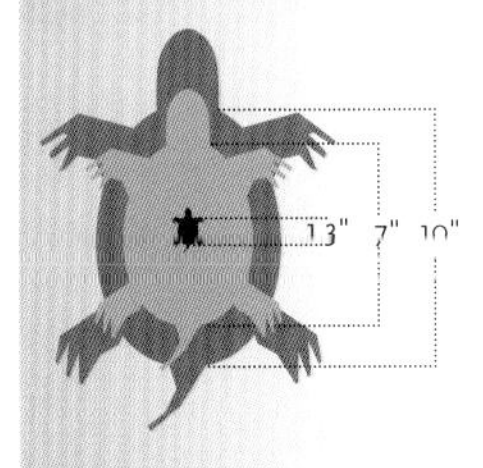

● FEMALE
● MALE
● HATCHLING

CARAPACE SHAPE

FRONT VIEW

SIDE VIEW

TOP VIEW

VARIATION AND TAXONOMIC ISSUES Three distinctive subspecies are recognized. The yellow-bellied slider turtle, *T. s. scripta,* has a dark shell with broad yellow bands visible on the sides of the carapace of young individuals and some adults; the front of the plastron generally has only two black spots; the broad patch behind the eye is yellow. The red-eared slider, *T. s. elegans,* has a greenish to brown shell mottled with black patterning, and a plastron with black smudges on each scute; the shell is less domed than that of *T. s. scripta,* and on some individuals the shell edges are less serrated; the prominent broad blotch behind the eye is red. In the Cumberland slider, *T. s. troosti,* the yellow head stripe behind each eye is only slightly broader than the other head stripes. The yellow-bellied and red-eared subspecies intergrade in western Georgia and eastern Alabama, and most individuals in that

The thin yellow to orange-red stripe behind the eye identifies the Cumberland slider subspecies.

The bright red spot behind the eye is the signature marking of the red-eared slider turtle.

region show characteristics of both subspecies, including having an orange blotch behind the eye.

Slider turtles are colorful and attractive as hatchlings.

WHAT DO THE HATCHLINGS LOOK LIKE? Hatchlings are generally shaped like the adults but are more green (red-eared subspecies) or yellow (yellow-bellied subspecies), rounder, have a more distinct keel on the carapace, and have the full suite of colors that become muted or lost in the adults.

CONFUSING SPECIES Chicken turtles also have vertical yellow lines on the rump, but the slider's posterior carapace edge is serrated while the chicken turtle's is smooth. Except for melanistic males, slider turtles can be distinguished from chicken turtles, all cooters, and most map turtles within their range by the large yellow to red blotch behind each eye.

DISTRIBUTION AND HABITAT Slider turtles are found in all or part of every southeastern state, but are not native to most of Virginia, the Blue Ridge Mountains, or the Florida peninsula. These habitat generalists thrive in virtually every aquatic habitat in the Southeast within their natural range, and in some outside it. They are most common in permanent swamps and ponds, relatively permanent riverine floodplain swamps and oxbow lakes, and slow-moving rivers. They flourish in riverine reservoirs with silt bottoms and abundant log debris and vegetation. They are good overland travelers and frequently colonize isolated seasonal wetlands, farm ponds, and natural lakes.

BEHAVIOR AND ACTIVITY Sliders are active throughout their southeastern range during all warm-weather months and on sunny or warm days even in winter. Their heads are often seen peeking above the water's surface, they are encountered crossing highways from spring to fall, and they bask in large numbers on any available object: logs, rocks, banks, and the surface of floating vegetation. As seasonal wetlands dry, slider turtles commonly move overland to other sources of water; they do not usually aestivate on land or bury themselves in the forest floor.

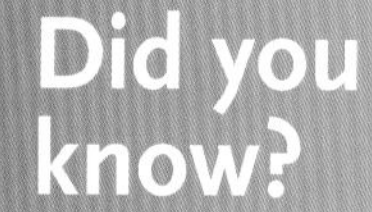

The red-eared slider probably has the largest geographic range of any freshwater turtle species in the world. However, only a small portion of that area is where it occurs naturally.

FOOD AND FEEDING Sliders are carnivorous (or insectivorous) as juveniles and omnivorous as adults; that is, adults can subsist on plant material alone but opportunistically eat any animal prey they can capture and scavenge on any dead fish or other animal they find. The list of plants and animals sliders eat is quite long, perhaps because of the extensive research that has been done on this species. It includes at least three kinds of algae; two dozen aquatic plants, including both native and introduced species; and many animal groups including freshwater sponges, snails, clams, crayfish, spiders, and numerous families of insects, fish, amphibians, and reptiles.

REPRODUCTION Like other species of turtles with elongated foreclaws, slider turtles go through a curious courtship ritual in which the male titillates the female by vibrating his claws in front of her face while both are under water. Courtship has been observed on sunny days in winter and early spring. Females nest in open, sunny areas near wetlands, often on roadsides and power line rights-of-way, and will travel more than a mile from water to find a suitable site.

Red-eared slider hatchlings were once sold by the millions in pet shops around the United States.

Adult male sliders often become melanistic, losing the yellow stripes on the head and markings on the shell.

The number of clutches per year varies from one to five, depending on the subspecies, and clutches may contain anywhere from 2 to more than 20 eggs (average = 6–11). Although the eggs hatch in late summer, hatchlings usually overwinter in the nest and emerge the following spring to travel to the water. Males reach maturity after as few as 2 or as many as 6 years, depending on environmental conditions. Females are usually not capable of laying eggs until they are at least 8 years old.

PREDATORS AND DEFENSE Nest predators include terrestrial mammals such as skunks and raccoons. Juveniles and adults fall prey to a host of native predators, including gar, catfish, snakes, alligators, crows, wading birds, raccoons, mink, otters, and coyotes. The largest U.S. specimens of the slider turtle came from three locations in South Carolina (Caper's Island and Kiawah Island near Charleston, and Par Pond Reservoir near Aiken) where alligators were the predominant large predators. Although large sliders captured in these sites often had bite marks and sometimes alligator teeth embedded in the carapace, their thick, high-domed shells apparently protected them from the awesome crushing power of the alligators' jaws.

The yellow-bellied and red-eared subspecies intergrade throughout the lower Gulf states. Most individuals in that region show characteristics of both subspecies, including having an orange blotch behind the eye.

CONSERVATION ISSUES Slider turtles present a major conservation puzzle. For decades, hatchling red-eared slider turtles were sold in five-and-dime stores throughout the United States, Europe, and elsewhere. Most of those hatchlings died, but those that survived were released by well-meaning people, and populations are now established in places far outside the native range. Red-eared sliders breeding with yellow-bellied sliders in eastern Virginia may change the character of local populations. They may hybridize with closely related slider turtle species outside the Southeast, even in other countries, and their genes may swamp unique traits of those species. They have been reported to displace native turtles in Europe and have become established in Southeast Asia with unknown consequences for that continent's ecosystem. Ironically, use of slider turtles for pets and food in China may actually help reduce pressure on Asia's threatened native turtle fauna. Clearly, some international cooperation is needed to address the conservation issues associated with the red-eared slider turtle.

An adult Florida cooter from North Carolina. The carapace of Florida cooters is taller and more rounded than that of river cooters.

How do you identify a pond cooter?

DISTINGUISHING CHARACTERS
Yellow unmarked plastron; yellow or greenish head stripes

AVERAGE SIZE

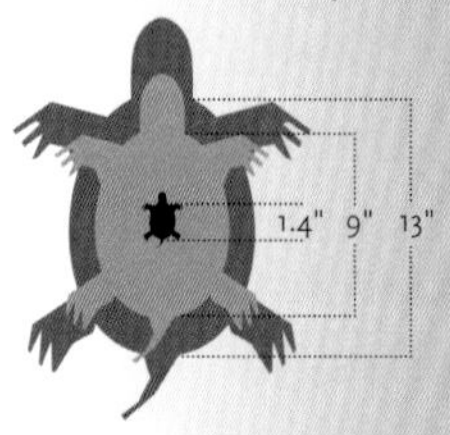

● FEMALE
● MALE
● HATCHLING

CARAPACE SHAPE

FRONT VIEW

SIDE VIEW

TOP VIEW

Pond Cooter

Pseudemys floridana

DESCRIPTION Pond cooters are large turtles with hard shells that are relatively high domed in front and taper off toward the rear. The carapace is brown to black with vague yellow markings, and the plastron is yellow with no dark markings. The marginal scutes are dark above with a touch of yellow below, typically with a dark spot with a light-colored center. The head stripes range from being prominent and yellow to faint and greenish. The hind feet are webbed. Females get appreciably larger than males, and mature males develop long foreclaws.

VARIATION AND TAXONOMIC ISSUES Two subspecies have traditionally been recognized, the more northerly one being called the Florida cooter, *P. f. floridana*, and the one found throughout the Florida peninsula known as the peninsular cooter, *P. f. peninsularis*. Some turtle biologists consider the peninsular cooter to be distinctive enough to warrant full species status (*Pseudemys peninsularis*), while others still consider it a subspecies of the pond cooter. See the river cooter (*Pseudemys concinna*) account for additional issues of geographic variation and taxonomy of this species.

WHAT DO THE HATCHLINGS LOOK LIKE? The carapace is greenish and has a keel down the center. The plastron is typically yellow and immaculate but with dark marks on the yellow underside of the marginals.

Pond Cooter
Pseudemys floridana

The subspecies found throughout the Florida peninsula is known as the peninsular cooter, *P. f. peninsularis*. Some turtle biologists have considered it a separate species (*Pseudemys peninsularis*).

The hatchling's carapace (left) is greenish and has a keel down the center. The plastron (right) is typically yellow and immaculate but with dark marks on the yellow underside of the marginals.

CONFUSING SPECIES The most effective way to differentiate this species from other hard-shelled turtles that occupy the same geographic range is by its yellow plastron with no dark markings and the dark, rounded marks beneath the yellow marginals that are light colored in the center. River cooters usually have a faint or prominent orange-and-black pattern on the yellow plastron and concentric partial lines or circles on the carapace scutes instead of relatively straight yellow lines. Florida red-bellied cooters have broad reddish stripes on the sides of the carapace.

DISTRIBUTION AND HABITAT The Florida cooter is found in the Coastal Plain from the Great Dismal Swamp in Virginia south to northern Florida and west to Mobile Bay, Alabama, generally in floodplain swamps and large lakes and ponds. It occurs in the slower-flowing sand- and silt-bottomed portions of Coastal Plain rivers, where it occasionally hybridizes with the river cooter. The latter species occupies the same rivers but is generally found in stretches above the Fall Line. The peninsular cooter is confined to Florida, where it inhabits large limestone spring runs, large sinkhole ponds and lakes, and the Everglades.

The peninsular cooter is confined to Florida, where it inhabits large limestone spring runs, large sinkhole ponds and lakes, and the Everglades.

BEHAVIOR AND ACTIVITY Like other turtles occupying a geographic range from the tip of Florida to Virginia, pond cooters hibernate during long cold spells but may be active year-round in southern regions. They are frequently observed basking on logs and limbs in the water but are usually wary and are much more difficult to approach than sliders or painted turtles. Pond

cooters that live in seasonal wetlands will travel overland to permanent water when their seasonal wetlands dry. Adults and juveniles of both sexes can be seen crossing roads.

FOOD AND FEEDING Hatchlings and juveniles are omnivorous, occasionally eating a variety of aquatic animals as well as plants, but adults are primarily herbivorous, with coontail, arrowhead, and duckweed being among the plants consumed.

REPRODUCTION Females mate in the water in spring and search for open, sunny areas, including roadsides and power line rights-of-way, to lay their eggs. Florida cooters generally nest in late spring and early summer. Peninsular cooters may nest year-round but most frequently nest from autumn through winter and into early spring. Even when eggs hatch in summer or early fall, the hatchlings may remain in the nest over the winter, emerging the following spring. Females may produce one to three or more clutches per year, with each clutch containing 10–30 eggs (average = 20). Peninsular cooters in Florida and Florida cooters in South Carolina may dig a main nest chamber and one or two adjacent "satellite" nests, into which one egg is usually deposited.

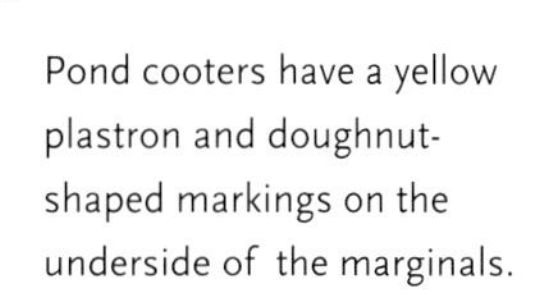

Pond cooters have a yellow plastron and doughnut-shaped markings on the underside of the marginals.

PREDATORS AND DEFENSE The known nest predators include black bears in addition to the usual suspects such as raccoons, skunks, and opossums. Small juveniles are preyed on by wading birds, snapping turtles, cottonmouths, and snail kites in Florida. Alligators are the primary threat to adult cooters, although the high-domed shells of adults may resist crushing by alligator jaws.

CONSERVATION ISSUES The Florida cooter and the peninsular cooter are frequently harvested for food. One approach to ensure sustainable local consumption would be to regulate the season when animals can be captured and the size and sex of individuals that can be taken, as is done with other wildlife game species; constant commercial exploitation may wipe out populations. Large basking turtles are often shot by local citizens who may not realize that these long-lived animals play an important role in controlling aquatic vegetation in many wetland systems.

Did you know?

Some cooters lay one or a few eggs in small nests (called satellite nests) they dig on each side of their main nest, which may contain more than a dozen eggs.

Florida red-bellied cooters are commonly seen basking.

How do you identify a Florida red-bellied cooter?

DISTINGUISHING CHARACTERS
Broad red bars on thick-shelled black carapace

AVERAGE SIZE

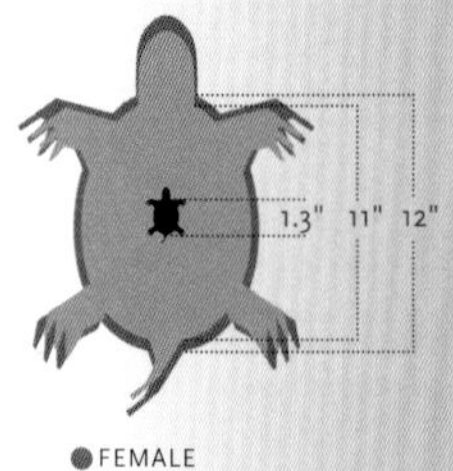

● FEMALE
● MALE
● HATCHLING

CARAPACE SHAPE

FRONT VIEW

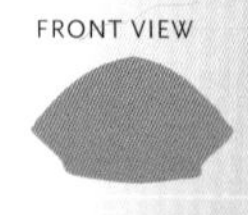

SIDE VIEW

TOP VIEW

Florida Red-bellied Cooter *Pseudemys nelsoni*

DESCRIPTION The Florida red-bellied cooter is a large, thick-shelled, high-domed turtle with a dark carapace and limbs made more colorful by the broad reddish bands on the sides of the carapace and the yellow stripes on the legs and on the sides and top of the head. The lines on the head are less conspicuous than in most other species of cooters, but one runs between the eyes to the nose, ending in an "arrow." The marginals have a hint of red above and dark markings below. The heavy plastron is reddish. The front claws of adult males are elongated. Females get larger than males, but the difference is not as striking as it is with the map turtles. Older, larger individuals become melanistic.

VARIATION AND TAXONOMIC ISSUES No subspecies or regional variation occurs in this species.

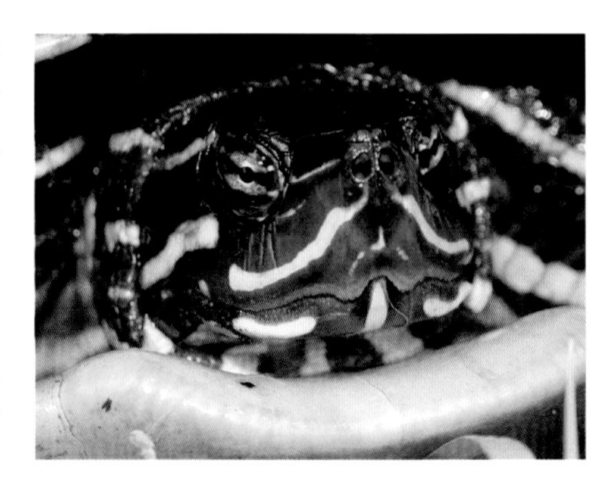

Florida red-bellied cooters often have fewer lines on the heads than other species of cooters.

WHAT DO THE HATCHLINGS LOOK LIKE? Young Florida red-bellied cooters are even more colorful than the adults, with a sharply patterned green-and-black carapace and a plastron that can be yellow but is more often orange to almost red.

CONFUSING SPECIES The range of this species falls within those of the closely related cooters and the slider turtle, which can be similar in appearance, but none of these have the red coloration prevalent in the Florida red-bellied cooter or the yellow "arrow" atop the head.

A hatchling.

DISTRIBUTION AND HABITAT The Florida red-bellied cooter's relatively small geographic range runs from the Okefenokee Swamp in Georgia to the Everglades, and is confined chiefly to the Florida peninsula. The species is abundant in the permanent open marsh and cypress swamp wetlands of the Okefenokee, occurs in sinkhole ponds throughout Florida, and is a common inhabitant of the many canals that traverse and parallel Florida's highways. Limestone springs and the clear, slow-moving, vegetation-choked small rivers that arise from them are also favorite habitats.

BEHAVIOR AND ACTIVITY Florida red-bellied cooters are active during daytime throughout the year in all but the most northern localities and are commonly seen basking. They are among the most noticeable basking turtles in the Okefenokee Swamp and along Florida canals. A noted propensity to bask on alligator nests presumably relates to their relative invulnerability

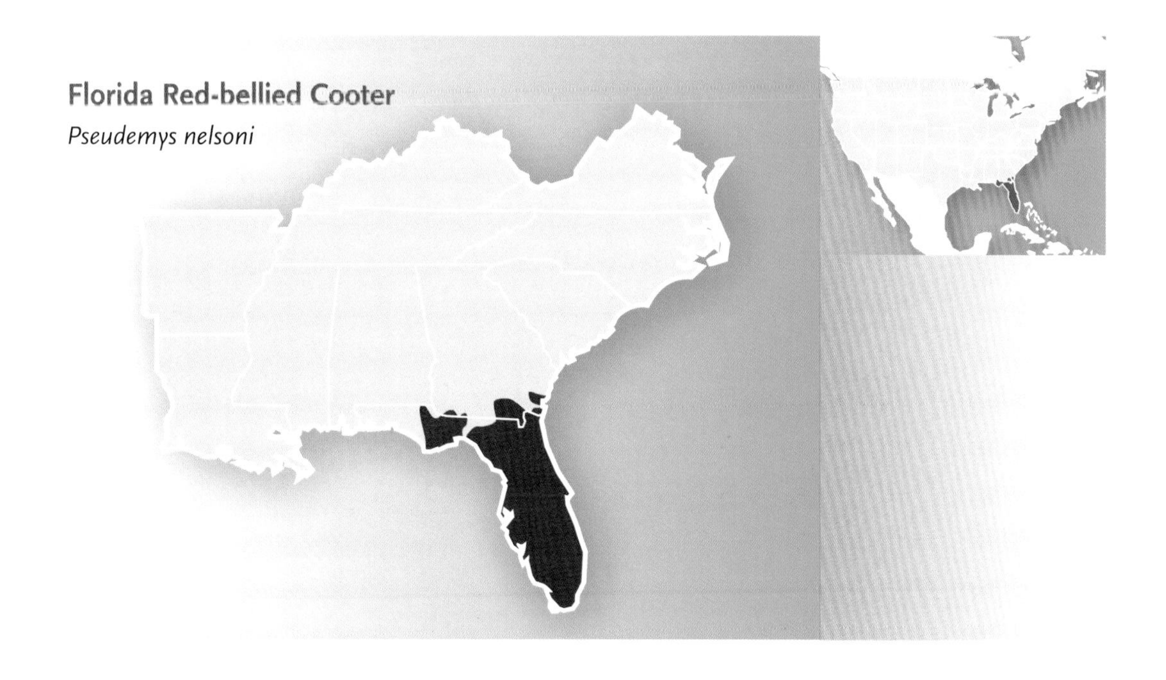

Florida red-bellied cooters are identified by the broad red stripes on the lateral sides of the carapace. They are common in Florida's karst springs.

to the giant predators. Snorkelers in Florida's big springs frequently see Florida red-bellied cooters.

FOOD AND FEEDING Adults are herbivores that eat algae and all sorts of aquatic vegetation, including duckweed, native pondweed, and at least two introduced plants: water hyacinths and hydrilla.

REPRODUCTION Florida red-bellied cooters are known to mate from late autumn to spring but presumably mate year-round. Mating is preceded by an elaborate underwater courtship in which the male titillates the female with his elongated claws. Nesting occurs during spring and summer, and perhaps throughout the year. Females lay from one to six clutches of 6–30 eggs, often depositing them within alligator nests. Female alligators aggressively protect their nests, and thus also the eggs of any cooter that has nested there.

PREDATORS AND DEFENSE Numerous native mammals are nest predators, including raccoons, otters, and skunks. In the aquatic habitat, the young turtles face an even greater array of predators, including large fish, cottonmouths, snapping turtles, and alligators. At least three predatory birds—bald eagles, ospreys, and snail kites—eat juveniles. The thick, high-domed shell of adults is likely an adaptation that allows them to coexist with alligators. Even adult alligators are typically unable to crack the shells of the largest of these turtles.

CONSERVATION ISSUES Most of this species' relatively small geographic range is in Florida, where human population levels and highway traffic densities are high. Road mortality of nesting females is therefore likely to become a major source of mortality for some populations.

Some northern red-bellied cooters are dark in coloration.

How do you identify a northern red-bellied cooter?

DISTINGUISHING CHARACTERS

Yellow arrow on head; plastron may be reddish

AVERAGE SIZE

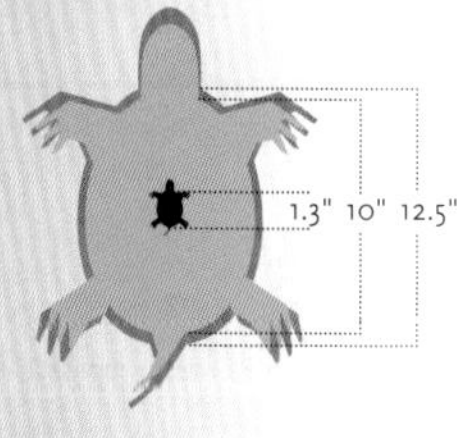

● FEMALE
● MALE
● HATCHLING

CARAPACE SHAPE

FRONT VIEW

SIDE VIEW

TOP VIEW

Northern Red-bellied Cooter

Pseudemys rubriventris

DESCRIPTION The northern red-bellied cooter is a large, dark brown to nearly black turtle with a high-domed shell. The carapace of younger individuals has a broad reddish or yellowish vertical line in each of the scutes on each side. The scutes around the margin are reddish above with a dark smudge on the underside. The black head and limbs have yellow pinstripes, with the central stripe on top of the head running between the eyes and ending at the nose in an arrow-shaped mark. The color of the hingeless plastron ranges from shades of yellow to orange to dull red. Juveniles and young adults often have a slight keel that runs down the center of the carapace. Adult males develop long front claws, and females become larger

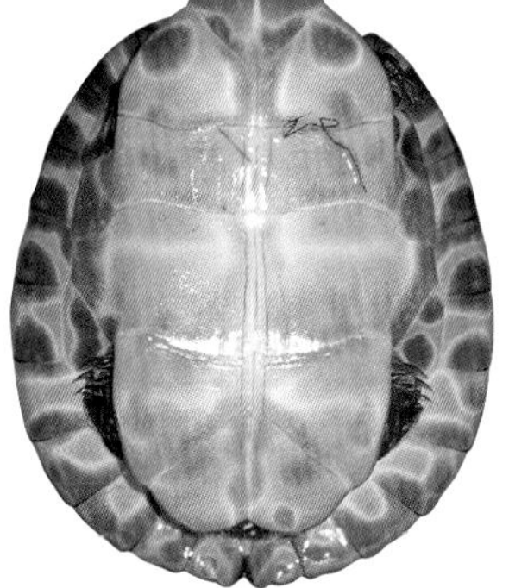

The red carapace markings (left) and plastron (right) of a juvenile.

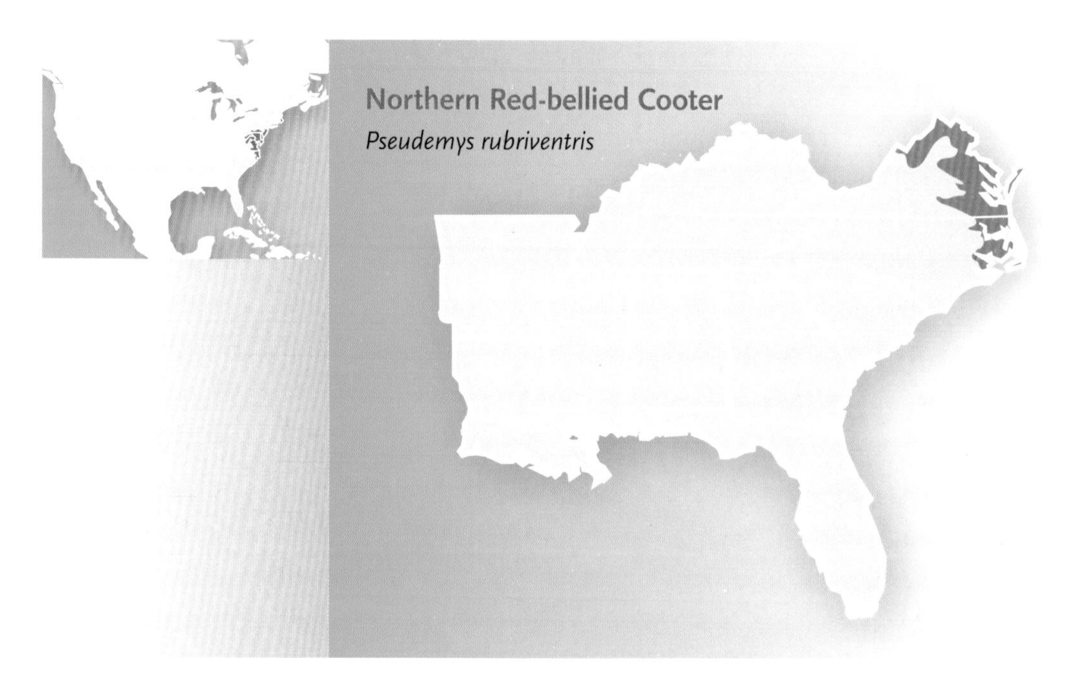

The hatchling's carapace (top) is greenish. A two-headed hatchling (bottom).

than males. The shell and limbs of large individuals of both sexes often become melanistic, which can diminish the red and yellow coloration and result in a mottled or reticulated pattern. Most adults have a small projection (cusp) on each side of the center of the upper jaw forming a notch at the midline.

VARIATION AND TAXONOMIC ISSUES No subspecies are formally recognized, and no geographic variation in color or pattern is apparent.

WHAT DO THE HATCHLINGS LOOK LIKE? Hatchlings are green above with a narrow yellow stripe or circle in each scute on the sides. The plastron is deep red with a black design that generally tracks along the seams.

CONFUSING SPECIES Other cooters, the slider, and the painted turtle are most likely to be mistaken for a northern red-bellied cooter in the southeastern portion of its range, but none of these species have the yellow "arrow" on the head. River cooters, pond cooters, and sliders also lack any red on the carapace.

DISTRIBUTION AND HABITAT In the Southeast, the northern red-bellied cooter is found in the mid-Atlantic Coastal Plain from northern Virginia to northeastern North Carolina, including a few locations on the Outer Banks. Large permanent wetlands, including beaver ponds, farm ponds, suburban ponds, and reservoirs as well as floodplain swamps, seem to be favored habitats, but these turtles are also found in slow-moving rivers.

BEHAVIOR AND ACTIVITY Because of its northern distribution, this species hibernates throughout its range, typically being active from March through October in the Southeast, although individuals may emerge to bask on warm, still days even during winter. Adults bask frequently throughout the year, and people are more likely to see one basking than engaged in any other activity.

The head pattern on northern red-bellied cooters is often dark.

FOOD AND FEEDING Adults are herbivores, primarily eating aquatic vegetation such as water lilies and pickerelweed. Juveniles are omnivores that eat insects, snails, crayfish, and tadpoles as well as aquatic vegetation.

REPRODUCTION Mating occurs from April through June. Males use their long claws during courtship, titillating the female with them as they swim backward in front of her or hover over her. Females nest in early to middle summer in open habitats, including agricultural fields and roadsides. Nesting typically begins during daylight hours, but the female may not complete the process until after dark. Females lay one or two clutches of 6–35 eggs (average = 10–12). The eggs hatch in late summer, but hatchlings sometimes spend the winter in the nest, emerging in the spring and moving overland to water.

PREDATORS AND DEFENSE Nest predation by raccoons is the major documented source of mortality, although crows, foxes, and skunks have also been identified as nest predators. Various aquatic animals such as fish and wading birds probably eat hatchlings and small juveniles; one known predator is the bullfrog. Other than nesting females that encounter raccoons on land, the adults presumably have few natural predators.

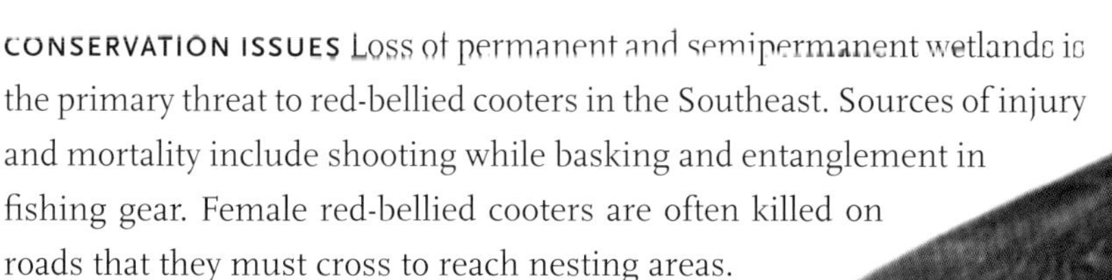

CONSERVATION ISSUES Loss of permanent and semipermanent wetlands is the primary threat to red-bellied cooters in the Southeast. Sources of injury and mortality include shooting while basking and entanglement in fishing gear. Female red-bellied cooters are often killed on roads that they must cross to reach nesting areas.

Most northern red-bellied cooters have obvious red bands on the sides of the carapace, as shown on this individual. However, some older individuals or those that live in black water streams have shells that are nearly black.

A young common snapper shows its long tail.

How do you identify a common snapping turtle?

DISTINGUISHING CHARACTERS
Pugnacious when threatened; long claws; long, serrated tail; reduced plastron

AVERAGE SIZE

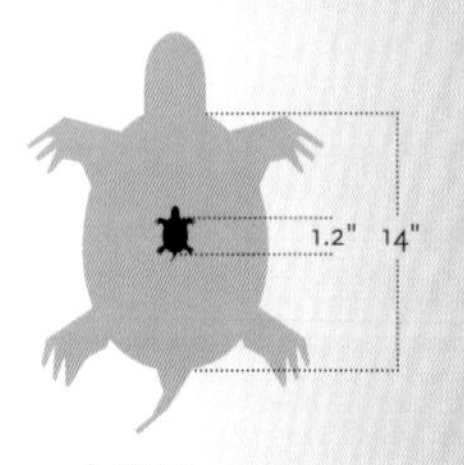

ADULT
HATCHLING

CARAPACE SHAPE

FRONT VIEW

SIDE VIEW

TOP VIEW

Common Snapping Turtle *Chelydra serpentina*

DESCRIPTION A full-grown snapper is the largest turtle a person is likely to see on land in the Southeast. They are frequently encountered crossing roads between bodies of water. Snapping turtles have dark gray to black shells and appendages of the same monotone color and are frequently infested with leeches. Their large head, strong jaws, sharp beak, and powerful claws give them a formidable appearance that is well deserved. The tail is long and is adorned with hard tubercles that give it a saw-toothed appearance. The carapace has a rough texture in younger animals but becomes smoother in large, old individuals, and its rear margin is deeply serrated. The plastron is greatly reduced compared with other southeastern turtles. The skin on the underside is lighter colored or creamy. Snappers derive their name from their tendency to snap or strike defensively when they feel threatened, and they are capable of delivering a painful bite. Male snappers are often larger than females, but the

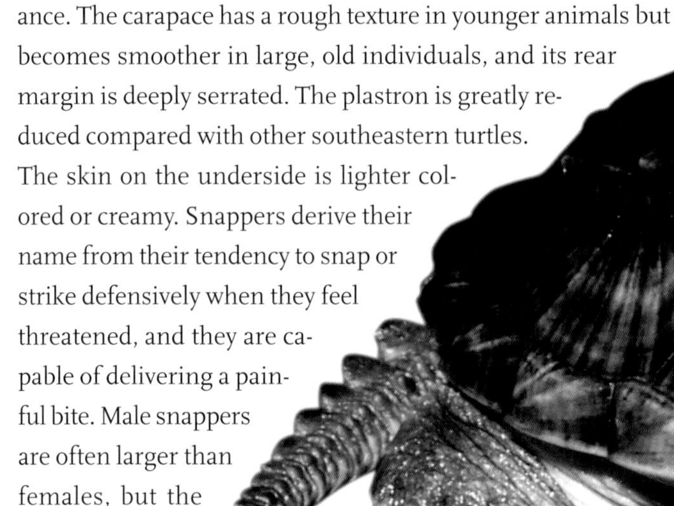

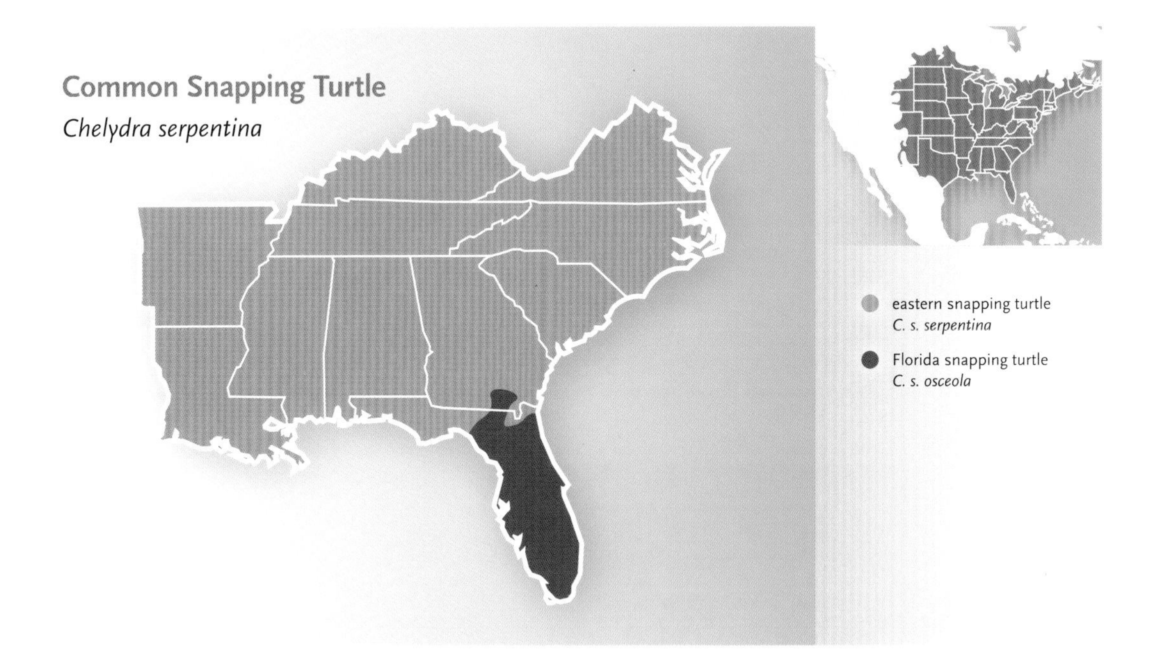

two sexes are otherwise indistinguishable except that the cloaca of adult males extends beyond the edge of the carapace.

VARIATION AND TAXONOMIC ISSUES Two of the four subspecies of the common snapping turtle occur in the Southeast: the eastern snapping turtle, *C. s. serpentina,* and the Florida snapping turtle, *C. s. osceola*. The differences between them are subtle, with the spines on the neck being slightly pointed in the Florida subspecies and more rounded and wartlike in the eastern subspecies. Some authorities question whether the Florida snapping turtle is distinct enough to warrant subspecies status. The most reliable determinant of a snapping turtle's subspecies is the geographic location where the specimen was found.

The formidable head of a common snapping turtle.

The eastern and Florida snapping turtle are almost indistinguishable if the location of capture is unknown.

The smiling face of a common snapping turtle.

WHAT DO THE HATCHLINGS LOOK LIKE? Babies look like rough-shelled versions of the adults, with a noticeably long tail and three intermittent keels down the length of the carapace. The plastron is darker than it is in older individuals and may be speckled with white spots. Each marginal scute has a distinctive white spot on the underside.

CONFUSING SPECIES A large common snapping turtle could be mistaken only for an alligator snapper, but the latter has a proportionately larger head and a hooked beak, and even adults have three distinct keels on the carapace. The keels disappear in large common snapping turtles. If close examination is possible, alligator snappers can be distinguished by the presence of an extra row of scutes between the marginal scutes and the costal scutes on the side of the carapace. The upper side of the alligator snapping turtle's tail is relatively smooth compared with the strongly serrated tail of the common snapper.

Hatchlings look like rough-shelled versions of the adults.

DISTRIBUTION AND HABITAT Common snapping turtles occur throughout every southeastern state, from the Coastal Plain to the mountains, and in almost every aquatic habitat, including rivers, farm ponds, reservoirs, oxbow lakes, floodplain swamps, and freshwater marshes. The Florida subspecies is restricted to the Florida peninsula.

BEHAVIOR AND ACTIVITY Snapping turtles are primarily aquatic, but both males and females can be found on land during the warmer months, traveling from one wetland to another or, in the case of females, to and from nesting areas. In the water, snappers are active both day and night, walking along the wetland bottom, plowing through thick vegetation, or swimming in open water. They frequently bury themselves in mud or hide under stumps and logs underwater. Snappers are active year-round in southern Florida but hibernate in the colder parts of their range. They can move slowly through almost freezing water and have even been observed poking their noses above the surface of partially frozen lakes. Common snappers seldom bask out of the water the way many other southeastern species do, although they often bask cryptically while floating on the surface among aquatic plants such as duckweed or water lilies.

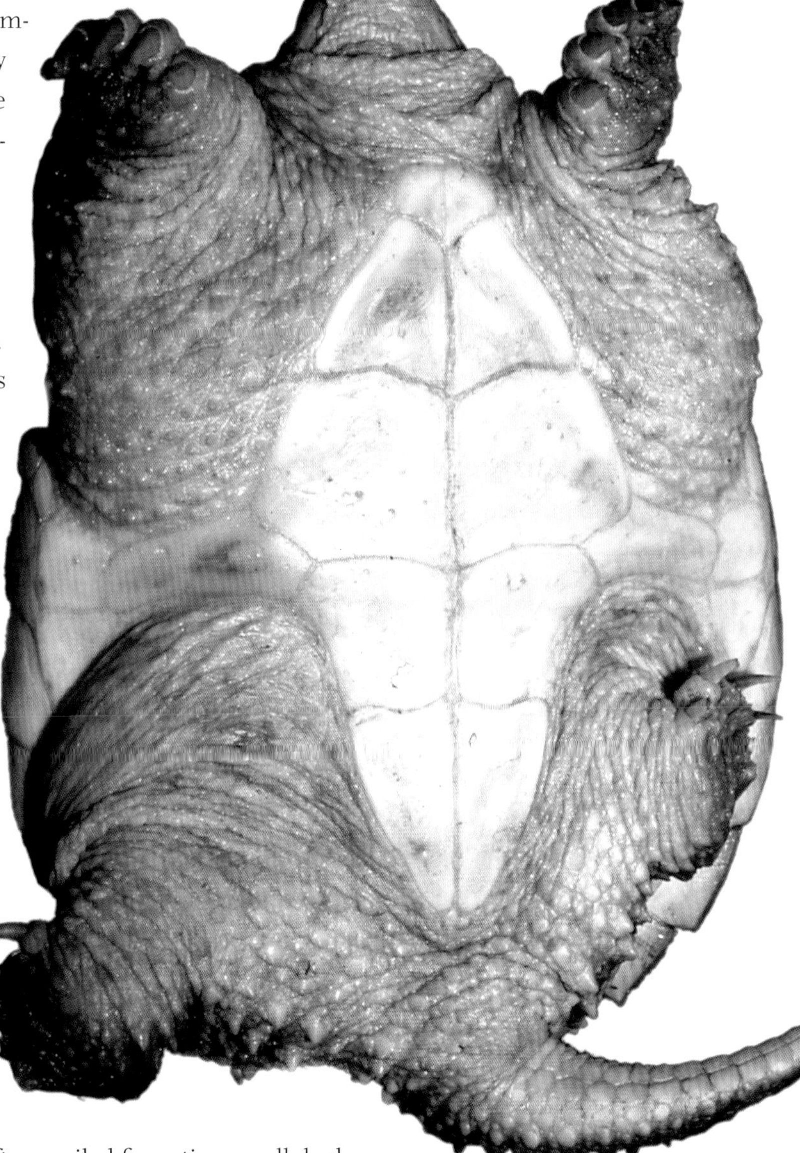

The plastron of the common snapper is reduced compared with other southeastern turtles.

FOOD AND FEEDING Common snappers are noted for eating a wide variety of both plants and animals—floating plants, insects, crayfish, other invertebrates, and carrion, for the most part—and for finding food both by active searching and by sitting in wait to ambush unsuspecting prey. They are also proficient scavengers. Snapping turtles are often reviled for eating small ducks and geese, but their reputation may be overstated. Unsuspecting, sick, or dead fish, frogs, salamanders, water snakes, other turtles, and birds also

Snapping turtles often walk overland with their plastron raised up off the ground.

fall prey to or are scavenged by snapping turtles.

REPRODUCTION Male snapping turtles are sometimes seen engaging in combat in the water, a tumultuous process that is presumably related to competition for mates. During the short springtime nesting period, female snappers search for open exposed areas to lay their eggs. Nests are often within a few feet of the water, but females have been recorded moving overland across roads and lawns from the water to nesting sites and back. Clutch sizes are the largest among the North American freshwater turtles. Females typically lay 20–40 eggs but may lay more than 100. The eggs hatch in late summer, and most hatchlings leave the nest soon afterward.

Because they bask while submerged in the shallows, common snappers frequently have algae on their shells and occasionally have it on their heads as well.

PREDATORS AND DEFENSE As with other turtles, the nests of snapping turtles are highly vulnerable to land predators such as raccoons, foxes, and skunks, which also eat hatchlings and small juveniles. Other predators of small turtles include large wading birds, hawks and eagles, carnivorous fish, and aquatic snakes. The stilted posture of an adult snapping turtle, with the hind legs elevating the rear of the shell, accompanied by a lunging up or forward with open mouth is familiar to anyone who has encountered a common snapper on land. It presumably is a defensive behavior that would discourage most animals. They also emit a foul-smelling musk when agitated. However, even large snappers can become prey of the largest alligators and, during winter dormancy, to river otters.

CONSERVATION ISSUES Common snapping turtles have experienced cyclical popularity in the restaurant trade, and in some areas they may be taken in unsustainable numbers. Nonetheless, they appear to be widespread and abundant in many areas of the Southeast. Wildlife management agencies should consider monitoring population status and harvest data, however, as this animal is an important component of nearly every aquatic ecosystem.

Common musk turtles are often covered with algae since they bask infrequently.

How do you identify a common musk turtle?

DISTINGUISHING CHARACTERS
Musky odor when handled; yellow stripes on head; reduced plastron with skin showing at seams

AVERAGE SIZE

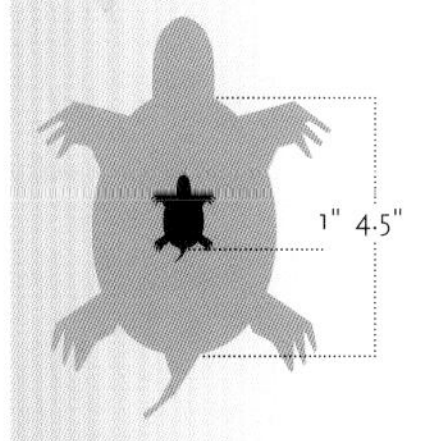

ADULT
HATCHLING

CARAPACE SHAPE

FRONT VIEW

SIDE VIEW

TOP VIEW

Common Musk Turtle or Stinkpot

Sternotherus odoratus

DESCRIPTION The common musk turtle is characterized by its brownish, nondescript shell and small, yellowish plastron with a single nonfunctional hinge. Skin shows between the plastral seams, and barbels are present not only on the chin but also on the neck. Most common musk turtles have a pair of pale or dark yellow stripes on each side of the head, and adults often retain a perceptible keel on the carapace midline. Two features occur commonly throughout the range: many individuals carry a mat of algae on part or all of the carapace, and the plastron and underside of the legs frequently bear leeches. Males have an impressively larger tail than females, and the tip of the male's tail ends in a blunt nail.

VARIATION AND TAXONOMIC ISSUES Despite its extensive geographic range, the common musk turtle has no recognized subspecies and little regional

Barbels are present not only on the chin but also on the neck.

variation. Some populations from streams and rivers in limestone areas with high mollusk diversity, such as the Cumberland Plateau, have wide heads, and this may be a local adaptation to an abundant food resource.

Hatchling common musk turtles are the smallest southeastern turtles.

WHAT DO THE HATCHLINGS LOOK LIKE? Common musk turtle hatchlings are the tiniest turtles in North America. They look like the adults except that the carapace has a distinct high keel down the center. The yellow lines on the head appear sharp and distinct.

CONFUSING SPECIES The mud turtles that overlap the range of this species have two hinges on the plastron rather than a single one, their carapace is never keeled, and no skin shows between the plastral scutes. The two yellow stripes on the head distinguish the common musk turtle from the other musk turtles, but striped mud turtles also have head stripes.

DISTRIBUTION AND HABITAT Common musk turtles are found throughout most of every southeastern state, being absent only from the higher elevations of the Blue Ridge Mountains. They are most at home in permanent wetlands such as ponds and lakes with abundant aquatic plants, but they also frequent streams, rivers, and swamps. They are less likely to inhabit seasonal wetlands that dry frequently.

BEHAVIOR AND ACTIVITY Common musk turtles are mostly aquatic, but they will wander about on land during rainstorms without ever getting too far from water. They will also bask, but rarely in full sunlight, being more

Two stripes on the head and small overall body size help identify the common musk turtle.

The male stinkpot (left) has more areas of fleshy exposure in the center of the plastron than the female does (right).

likely to climb out of the water to sit on overhanging branches or at the fork of a limb. They often drop with a *plop* into the water or even a boat that happens to be passing underneath. They are poor overland travelers and are unlikely to move from one wetland to another as most other semi-aquatic turtles do. The empty shells of common musk turtles that turn up around small wetlands suggest that they are among the most obvious casualties during droughts. Common musk turtles are called "stinkpots" in some regions because of the presence of glands between the carapace and plastron that ooze an orange substance that is particularly foul smelling. Presumably this deters some predators from eating them, and it may also play a role in mate attraction. Musk turtles are adept at biting the hand

Musk turtles rarely travel far from water.

that holds them, as they can turn their head unexpectedly far behind them.

FOOD AND FEEDING Common musk turtles will apparently eat any type of aquatic plant or animal, living or dead. They have been observed to feed on algae, water lilies, water hyacinths, insects, crayfish worms, leeches, snails, and dead fish and mollusks.

REPRODUCTION Common musk turtles probably mate throughout the year in the Southeast, with breeding peaks in the spring and fall. Females lay from one to four clutches of 1–5 eggs each, nesting at night as well as during the day, usually within 150 feet of the water's edge, often in shady areas. In contrast to the characteristic flask-shaped nest females of most other southeastern turtle species dig in the soil, common musk turtles often lay their eggs beneath logs, under leaves, or even on open ground. Some hatchlings in the Southeast overwinter in the nest and arrive at the water in spring, but most emerge from the nest when they hatch in late summer or fall.

Common musk turtles will occasionally climb several feet to bask on overhanging branches or protruding limbs.

PREDATORS AND DEFENSE Common musk turtle eggs are susceptible to numerous terrestrial predators, including kingsnakes and scarlet snakes, raccoons, and crows. Juveniles and adults are eaten by an array of larger predators, including red-shouldered hawks, bald eagles, largemouth bass, bullfrogs, cottonmouths, river otters, and perhaps boat-tailed grackles. Adults are slow moving both on land and in the water, and will try to bite predators and emit an obnoxious fluid.

CONSERVATION ISSUES Common musk turtles seem to be abundant in most places where they occur. Because they rarely venture far from water, they are less vulnerable to road mortality than most of our other turtles. The common musk turtle will likely be one of the last turtle species to be negatively affected by environmental degradation or by unsustainable collection for any purpose.

The leathery carapace of this female has rounded tubercles on the front edge.

Florida Softshell Turtle *Apalone ferox*

DESCRIPTION Florida softshell turtles get longer and heavier than other species of North American softshells, with females growing twice as large as males. The leathery carapace has rounded tubercles on the front edge and an almost rough appearance. The limbs are gray to brown with light mottling. Like other softshells, this species has webbed feet suitable for its almost fully aquatic existence and an extended nose tip. Large males and females develop jaws and heads that are disproportionately enlarged relative to their body size.

Hatchlings are much more brightly colored than adults.

VARIATION AND TAXONOMIC ISSUES No subspecies or geographic variation has been described for this species.

WHAT DO THE HATCHLINGS LOOK LIKE? Florida softshell hatchlings are among the prettiest and most distinctive turtles in North America and are unlikely to be confused with any other species. They are readily distinguishable from the other softshells by their dark head with bright yellow and orange markings, dusky plastron, and olive carapace with a light border and dark blotches.

How do you identify a Florida softshell turtle?

DISTINGUISHING CHARACTERS
Elongate, leathery shell; snorkel-like snout; rounded bumps on front of carapace

AVERAGE SIZE

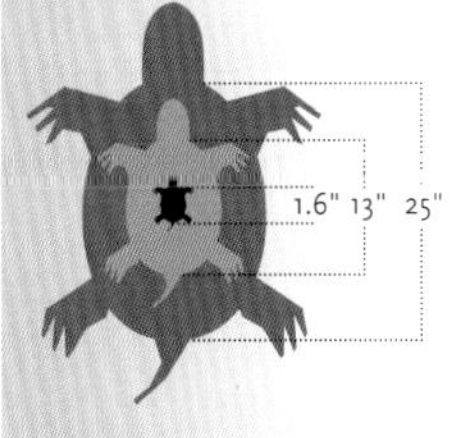

FEMALE
MALE
HATCHLING

CARAPACE SHAPE

FRONT VIEW

SIDE VIEW

TOP VIEW

Prominent striping is visible on the head in hatchlings.

CONFUSING SPECIES Adult Florida softshell turtles are easily distinguished from the other two species of softshells in the Southeast by their darker, more oval shell, which also has a slightly upturned rim. Like the spiny softshell, but unlike the smooth softshell, a tiny but obvious ridge is visible inside the nostril. Small bumps on the surface of the front part of the carapace are not as pointed as those on the shell of the spiny softshell.

A top view shows the elongate carapace and rounded tubercles.

DISTRIBUTION AND HABITAT Florida softshell turtles are found throughout Florida, the Coastal Plain of Georgia, extreme southern Alabama, and southeastern South Carolina. They are at home in relatively permanent waters such as large sinkhole ponds, limestone springs, and floodplain swamps, as well as slow-moving rivers and canals with silty bottoms and aquatic vegetation, but they are primarily pond dwellers.

BEHAVIOR AND ACTIVITY Florida softshells are active in the water during all months of the year in southern Florida and during all but the coldest periods of winter in other areas. They apparently are capable of underwater respiration through their skin, or possibly through tissues in the throat. Although primarily aquatic, Florida softshells occasionally venture overland between bodies of water.

FOOD AND FEEDING Small to medium-sized Florida softshells primarily eat aquatic insects and snails. Larger individuals increase the amount of fish in the diet and take large mollusks and crayfish. Older individuals may be able to eat a wider variety of prey because of their disproportionately larger jaws, which are capable of subduing larger prey.

REPRODUCTION Courtship and mating patterns in Florida softshells are not well known. Females lay two or more clutches of 4–24 eggs. Nesting females have been observed to disturb the ground several feet away from a recently covered nest, although the purpose of this behavior is unknown. Nesting begins in mid-March in Florida and in early summer farther north in the range.

The lateral ridge is visible inside the nostril of the spiny softshell and this Florida softshell.

PREDATORS AND DEFENSE Florida softshell nests are subject to a variety of egg predators such as raccoons, skunks, foxes, and crows. Hatchlings are eaten by other turtles, fish, wading birds, and even snail kites. Adults, especially large females, are less vulnerable, but even they can fall victim to alligators, which occur throughout the species' range. Camouflage and concealment on the bottom are primary defenses, and large individuals can deliver a serious and painful bite.

CONSERVATION ISSUES This species is frequently sold commercially for food; hence efforts must be made, especially in Florida, to ensure sustainable removal. Hatchlings are exported in high numbers; most go to China, where the release of captives into the wild could result in their becoming established there. Females are often killed while crossing roads during the nesting season. Florida softshells are among the most likely turtle species to go after baited hooks and are killed routinely by fishermen who cut the turtle's head off to get their hook back. They are caught both intentionally and incidentally on trotlines.

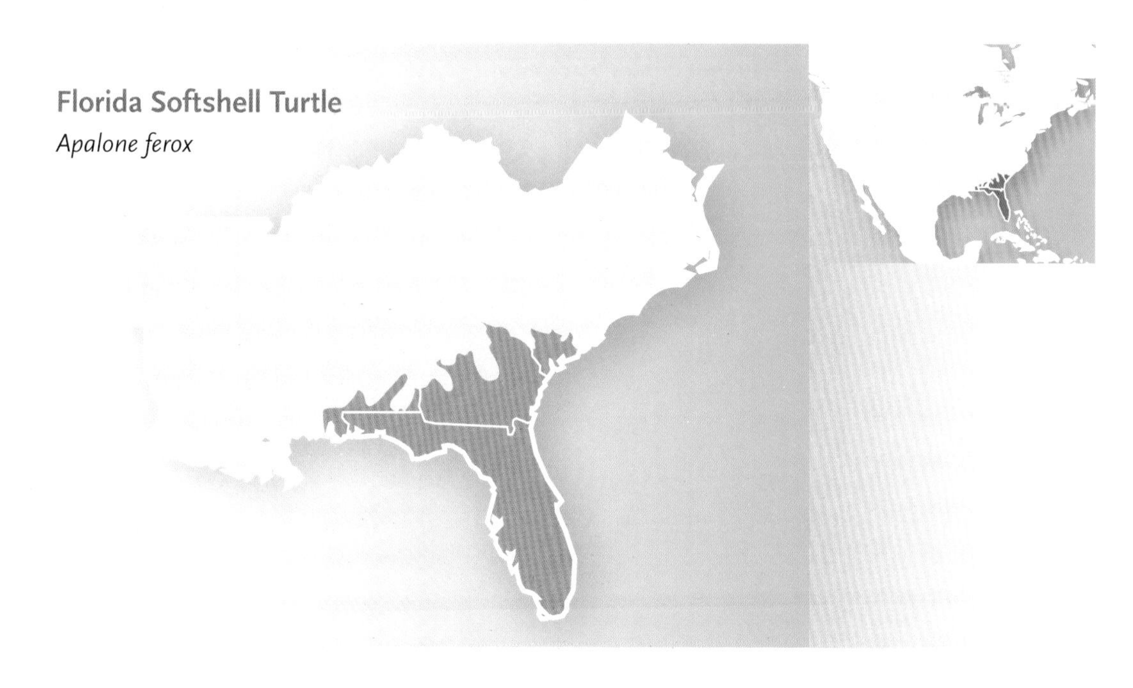

RIVERINE TURTLES

The alligator snapper has an enormous head, a hooked beak, and three ridges on the carapace.

How do you identify an alligator snapping turtle?

DISTINGUISHING CHARACTERS
Very large with massive head; three ridges down carapace; long smooth tail

AVERAGE SIZE

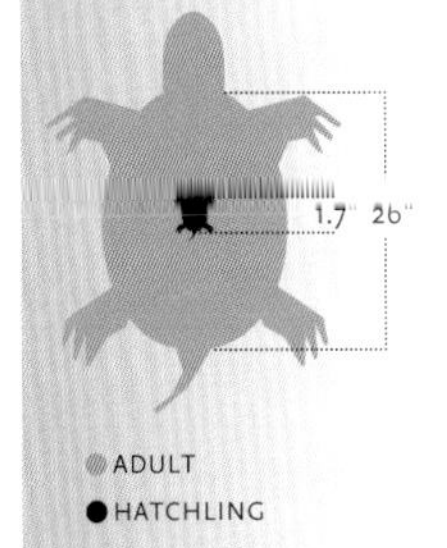

● ADULT
● HATCHLING

CARAPACE SHAPE

FRONT VIEW

SIDE VIEW

TOP VIEW

Alligator Snapping Turtle *Macrochelys temminckii*

DESCRIPTION The alligator snapper, with its massive head ending in a strongly hooked upper jaw, is the largest freshwater turtle in North America. The tail is long and may have rounded bumps but is not saw-toothed. The dark brown carapace has three longitudinal keels and is deeply serrated in the rear. The plastron looks disproportionately small compared with that of most other turtles. Males can be enormous, sometimes almost doubling females in both length and weight. Some large, presumably old, males have white or yellowish heads.

VARIATION AND TAXONOMIC ISSUES No subspecies have been described, and no significant geographic variation has been observed.

WHAT DO THE HATCHLINGS LOOK LIKE? Hatchlings are similar in appearance to the adults, but with the hook on the upper and lower beaks, the serrations along the rear margin of the carapace, and the three keels down the center and sides of the carapace more pronounced.

Hatchlings, similar in appearance to the adults, also display the wormlike lure.

The distinct ridges (above) identify an alligator snapping turtle. This male weighs 97 pounds.

The pink wormlike lure on the bottom of the mouth (right) is used to trick unwary fish into entering the alligator snapper's open mouth.

CONFUSING SPECIES Small alligator snappers differ from musk turtles in having a relatively small, hingeless plastron; long tail; and rough carapace. They can be distinguished from common snappers by the three distinct keels that are retained on the carapace, the row of supramarginal scutes between the marginal scutes and the costal scutes on the side of the carapace, the larger head, and the relatively smooth surface of the upper side of the tail compared with the strongly serrated tail of common snappers. Hatchling and juvenile alligator snapping turtles lack the white spots that edge the carapace of common snapper hatchlings. The inside of the mouth is mottled gray-brown with black speckles, while in common snappers the mouth is white or pinkish.

The plastron is greatly reduced on alligator snapping turtles.

DISTRIBUTION AND HABITAT The alligator snapping turtle is restricted to river and stream drainages that flow into the Gulf of Mexico. Reports of alligator snappers from Atlantic drainages are probably common snapping turtles, although turtles are sometimes transported and released outside their known ranges. They are found in floodplain swamps and oxbow lakes associated with large rivers but do not occur in isolated wetlands and ponds. In the Southeast, the range of the alligator snapping turtle encompasses the Gulf Coastal Plain from the Suwannee River drainage of the Okefenokee Swamp through south Georgia and north Florida westward to Arkansas, Louisiana, and east Texas.

BEHAVIOR AND ACTIVITY This species is one of America's most aquatic turtles, almost never leaving the water except to nest. Individuals have

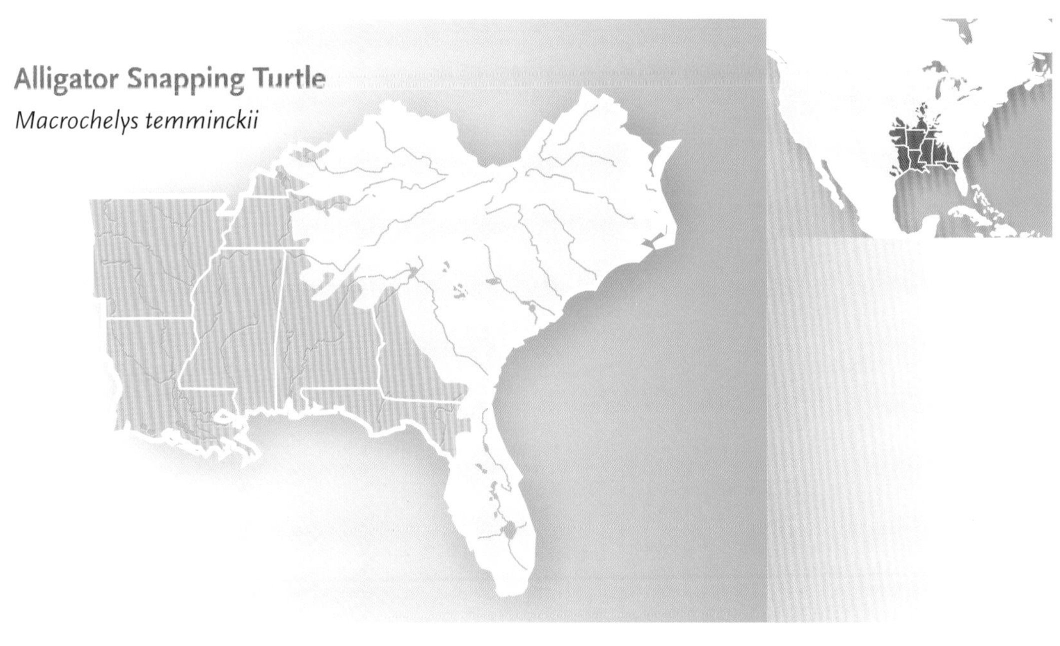

been recorded to move several miles both upstream and downstream in rivers. Male alligator snappers are aggressive toward each other and may maintain some territorial boundaries, possibly to control access to mates or feeding areas. Alligator snapping turtles are most active after dark, when they presumably search for food.

FOOD AND FEEDING Alligator snappers have a surprisingly diverse diet that includes plant material ranging from acorns and hickory nuts to aquatic grasses and berries, and animal prey ranging from aquatic snakes, turtles (especially musk turtles), and salamanders to large fish, small alligators, muskrats, and beavers. In clear waters, alligator snappers use their wormlike tongue as a lure, enticing fish and possibly small turtles into the enormous open mouth, which can snap shut with great speed and force. Juveniles may actually spend more time luring than adults, which may scavenge dead fish and consume acorns instead.

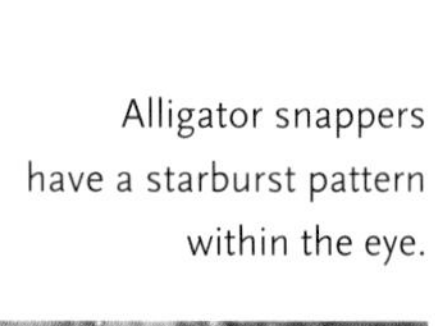

Alligator snappers have a starburst pattern within the eye.

REPRODUCTION Mating occurs in the water in late winter and early spring. During late spring or early summer females climb the elevated river sandbank to dig their nest. Construction often begins before dawn on warm, humid spring days following a rain and may take several hours. Tracks left

The alligator snapper has a massive head ending in a strongly hooked upper jaw.

The three carapace ridges are prominent on this juvenile. Also notice the hooked upper beak.

by one of these turtles lumbering up onto the bank resemble those made by a small bulldozer. One clutch of about 10–60 eggs is laid each year. Incubation time ranges from 80 to 114 days.

PREDATORS AND DEFENSE Mammalian predators and crows that eat the eggs and hatchlings in the nest and large fish, wading birds, snakes, and alligators that eat the juveniles in the water are the greatest sources of natural mortality. Adult alligator snappers can protect themselves from any natural predator with a single defensive bite. Hatchlings conceal themselves among underwater logs, sticks, and branches, or may bury themselves in silt along the edges of muddy sandbars.

CONSERVATION ISSUES Commercial hunters supplying the turtle soup trade nearly extirpated populations of alligator snapping turtles from numerous rivers and streams in the Southeast in the 1960s and 1970s. Trappers would abandon an area once the return became uneconomical, leaving few individuals behind. State laws in most states eventually regulated the take, initiating the species' recovery in some river systems. Full recovery to former population sizes, however, is predicted to take a long time for this long-lived, slow-to-mature species. Alligator snapping turtles are still vulnerable to capture on trotlines.

The razor-back musk turtle's carapace is steeply sloped with conspicuous, overlapping vertebral scutes.

How do you identify a razor-back musk turtle?

DISTINGUISHING CHARACTERS
Tent-shaped brown carapace; barbels on chin

AVERAGE SIZE

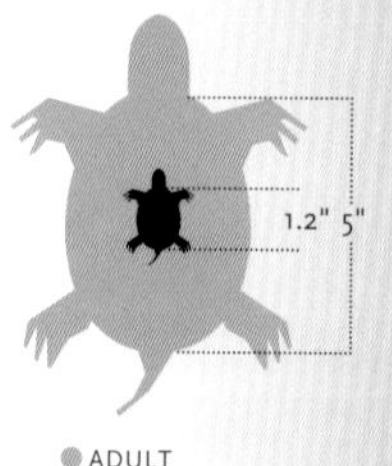

● ADULT
● HATCHLING

CARAPACE SHAPE

FRONT VIEW

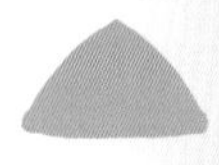

SIDE VIEW

TOP VIEW

Razor-back Musk Turtle *Sternotherus carinatus*

DESCRIPTION The razor-back musk turtle has a distinctly slanted, steeply sloped carapace that is tan to orangish brown with irregular black markings on all scutes. The markings may become obscure in larger individuals. The vertebral scutes on the carapace overlap from front to back. The reduced plastron is yellow with no markings and has an anterior hinge. The grayish head has numerous black specks. All four feet are webbed, and a pair of barbels is located under the chin.

VARIATION AND TAXONOMIC ISSUES No subspecies have been described, and regional variation is not apparent.

WHAT DO THE HATCHLINGS LOOK LIKE? Hatchlings look like light-colored versions of the adults, with more sharply overlapping scutes on the carapace. The plastron is pale with a hint of orange.

CONFUSING SPECIES No other musk turtles have such angular shells, and mud turtles have two plastral hinges instead of one. Small snapping turtles have a much longer tail and their plastron is hingeless.

DISTRIBUTION AND HABITAT Within the Southeast the razor-back musk turtle ranges from the Escatawpa River in Alabama and Pascagoula River in Mississippi west through Louisiana and much of Arkansas, including

Razor-back Musk Turtle

Sternotherus carinatus

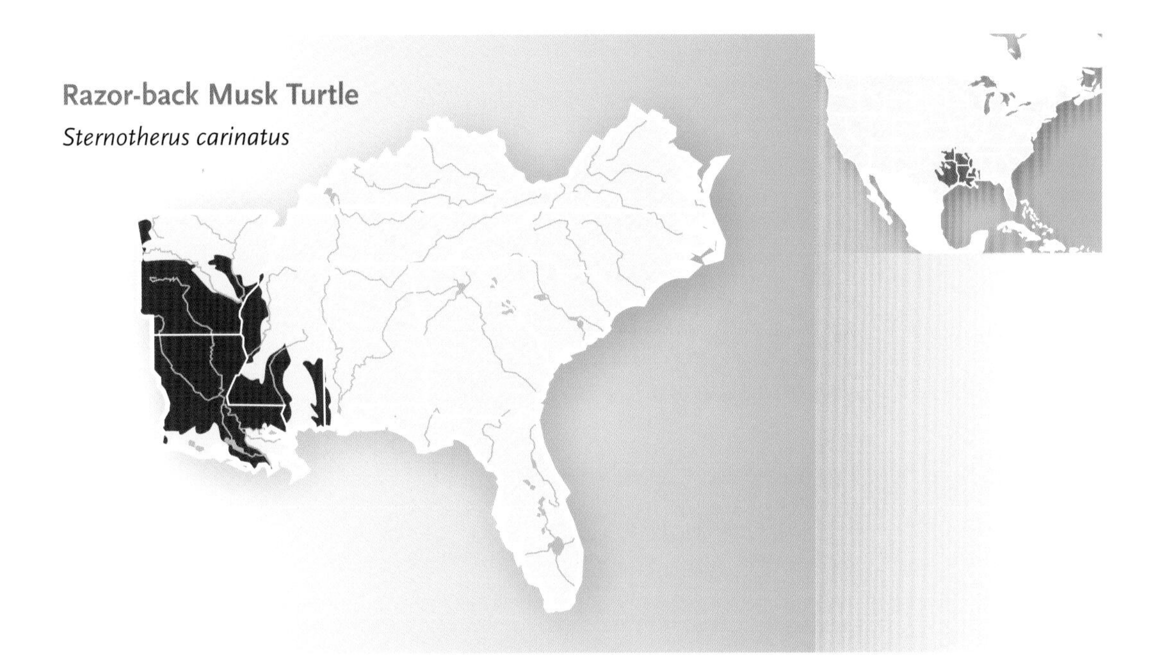

Hatchlings look like light-colored versions of the adults.

streams in the Ouachita Mountains. It inhabits rivers and streams with pools of deep water and substantial current.

BEHAVIOR AND ACTIVITY This bottom walker probably uses the barbels on its chin to "feel" for food in murky waters. Razor-back musk turtles differ in behavior from other musk turtles in that they seldom bite if picked up, do not excrete a foul musk, and are frequently seen basking. They are excellent climbers and are often observed on low-hanging tree branches over water. Aside from females on nesting excursions, they rarely travel on land. They are active throughout much of the year and spend only a short portion of the winter hibernating under rocks on the bottom of the stream or river or under the banks.

Carapace markings become obscure in larger and older individuals.

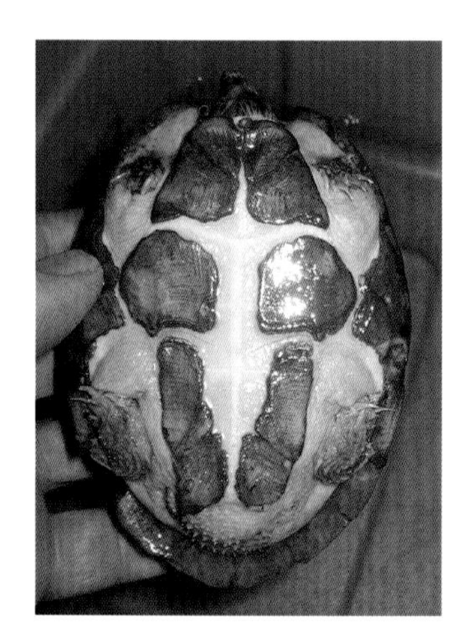

The reduced plastron has an inconspicuous anterior hinge. Shown here is an adult female.

FOOD AND FEEDING Razor-back musk turtles eat primarily animal material, especially snails and mussels but also live insects, crayfish, and tadpoles. They will eat algae and aquatic vegetation as well and will scavenge on dead fish.

REPRODUCTION Mating occurs in the water, and females nest along riverbanks. Females lay one or two clutches of 2–5 eggs in May and June. Little else is known about reproduction in this species; observations of natural nests are rare.

PREDATORS AND DEFENSE Presumably, the same nest predators that eat the eggs of other southeastern turtles prey on the eggs of this species. Predators of hatchlings include fish and water snakes. Juveniles and adults may fall prey to large catfish, alligator snapping turtles, and river otters. These are among the most inoffensive of the musk turtles. Most individuals are shy and seldom release musk.

CONSERVATION ISSUES Being restricted to rivers over a relatively small geographic range, the razor-back musk turtle is susceptible to pollution or sedimentation that could reduce or eliminate the aquatic insects and mollusks on which it feeds.

A loggerhead musk turtle from a clear spring run in Georgia.

How do you identify a loggerhead or stripe-necked musk turtle?

DISTINGUISHING CHARACTERS
Enlarged head; keeled carapace; head spotted in loggerhead, striped in stripe-necked

AVERAGE SIZE

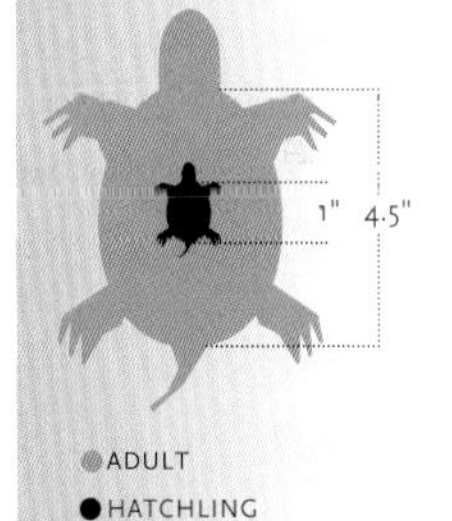

● ADULT
● HATCHLING

CARAPACE SHAPE

FRONT VIEW

SIDE VIEW

TOP VIEW

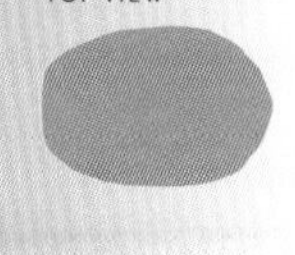

Loggerhead and Stripe-necked Musk Turtles

Sternotherus minor

DESCRIPTION The loggerhead (*S. m. minor*) and stripe-necked (*S. m. peltifer*) musk turtles represent two distinct subspecies of a species with a tan to brown carapace with black streaking and overlapping scutes down the center of the carapace. The carapace of younger turtles has three keels, but the keels are either absent or visible only on the rear of the carapace of old adults. The plastron has a single, rather inconspicuous hinge. Both sexes have a pair of barbels on the chin, and the male's tail has a curved, clawlike tip that he presumably uses to help position himself during mating.

VARIATION AND TAXONOMIC ISSUES The regional variation observed is mainly associated with the two subspecies, including a broad range of overlap and intergradation between them. Loggerhead musk turtles have a large brown or gray head with many small black spots and are more likely to retain remnants of all three carapace keels as adults. The head of old loggerhead musk turtles becomes disproportionately enlarged, as does the front half of the entire shell. The stripe-necked musk turtle has yellowish, broken stripes on the head and neck. The plastron and underside edges of the carapace are pale pink, yellow, or white in the loggerhead subspecies but orangish in the stripe-necked subspecies. Adult stripe-necked musk

Loggerhead and Stripe-necked Musk Turtle

Sternotherus minor

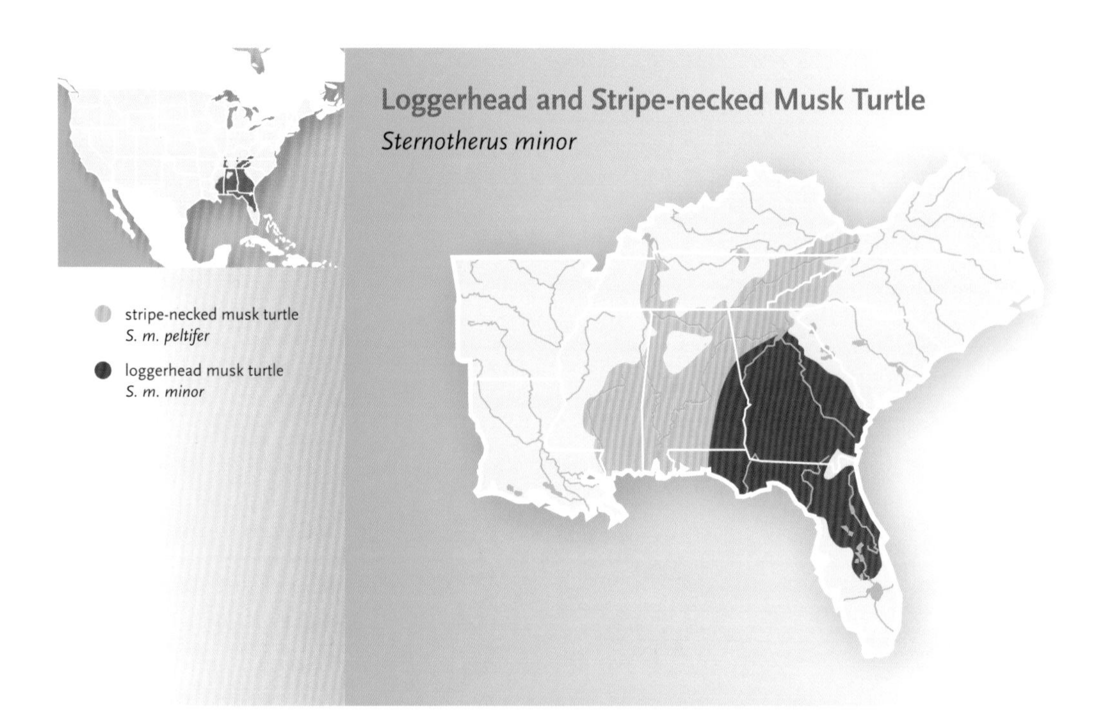

A hatchling loggerhead musk turtle shows part of its pink plastron.

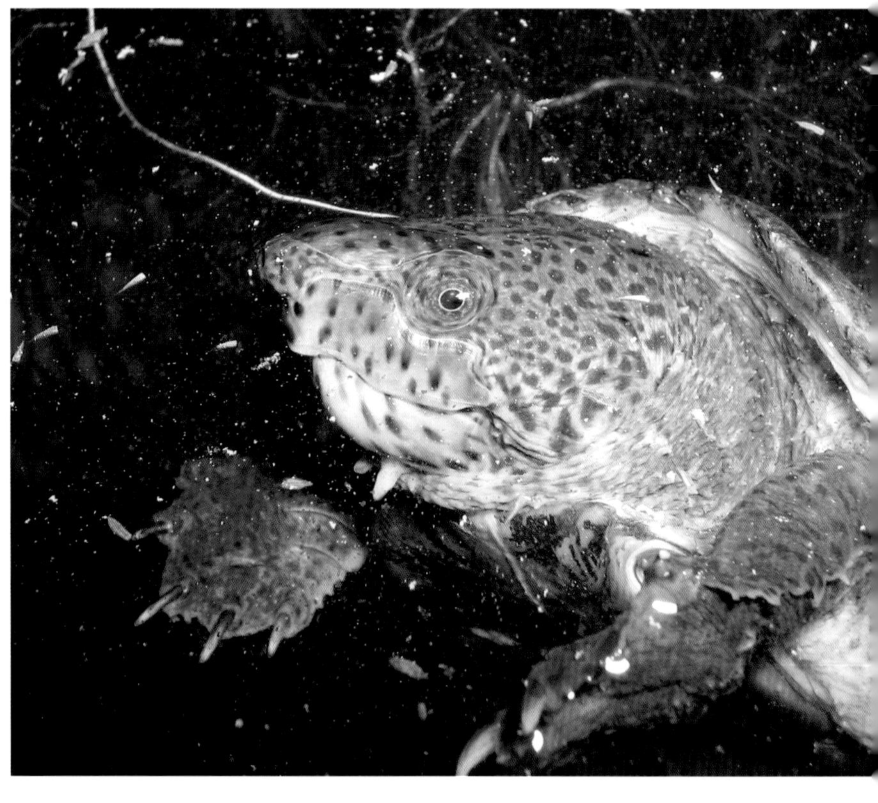

Loggerhead musk turtles have a large brown or gray head with many small black spots.

A juvenile stripe-necked musk turtle (left) from a stream in northern Alabama. The head striping is characteristic of turtles from the Cumberland Plateau and Piedmont regions.

A hatchling stripe-necked musk turtle with a tan carapace (top) and orange plastron (bottom).

turtles lack the keels on either side of the carapace, retaining only the single ridge down the middle of the back.

WHAT DO THE HATCHLINGS LOOK LIKE? The hatchlings resemble the adults but with more pronounced central and side keels. Hatchlings of the stripe-necked subspecies may have an orange plastron and only a trace of the keels on the sides of the carapace when compared with the loggerhead subspecies. The plastron of loggerhead musk hatchlings is bright pink.

CONFUSING SPECIES The other musk turtles within its range are the most likely to be confused with this species. The razor-back musk turtle has a more steeply sloped carapace, the flattened musk turtle has a greatly depressed shell, and the common musk turtle has yellow stripes on the sides of the head. The two species of mud turtles that overlap with the range of loggerhead and stripe-necked musk turtles in the Southeast have two distinct hinges on the plastron.

DISTRIBUTION AND HABITAT Loggerhead and stripe-necked musk turtles are found over a broad geographic area from central Florida to Louisiana to Virginia. The species is absent from the northern Alabama region occupied by the flattened musk turtle. The loggerhead musk turtle is a common resident of clear limestone springs in Florida and rivers and stream tributaries of the Gulf Coastal Plain and Piedmont regions of Georgia. Stripe-necked

The chin barbels are used to "feel" for prey in dark or turbid conditions.

musk turtles are characteristically found in rivers and streams of the Cumberland Plateau region, especially faster-flowing streams with rock and cobble bottoms.

BEHAVIOR AND ACTIVITY Except for nesting females, these musk turtles almost never venture out onto land. Loggerhead musk turtles can often be seen crawling on the bottoms of Florida springs both day and night at depths of 20 feet or more. They apparently can acquire oxygen from the water by absorbing it through the skin. They are also sometimes observed basking high above the water atop cypress knees or on the limbs of small trees. Less is known about the behavior of stripe-necked musk turtles, although they presumably spend much of their time on stream bottoms in search of prey.

FOOD AND FEEDING Both loggerhead and stripe-necked musk turtles eat aquatic insects, crayfish, and aquatic plants, and especially prefer small snails. They will scrape algae from rocks and may occasionally eat carrion. Larger individuals, especially old males, develop heavy jaws that help them crush snails and mussels.

REPRODUCTION Adults mate underwater in both spring and fall, often with more than one male pursuing a single female. Females lay two or more clutches of 1–5 eggs. Aside from a few anecdotal observations, little is known of the nesting habits or other reproductive behavior of this species.

PREDATORS AND DEFENSE Little is known about nest predation in this species anywhere in its range, but presumably raccoons, skunks, and crows take eggs from the nests. Alligators, snapping turtles, cottonmouths, and large fish that inhabit stream systems where loggerhead and stripe-necked musk turtles are found are presumably predators, especially of juveniles.

CONSERVATION ISSUES These turtles are susceptible to pollution of the rivers, streams, and springs in which they live. Sediments, chemicals, coal-mining waste, and road runoff harm populations of the mollusks they eat. Anecdotal observations indicate that the stripe-necked musk turtle has declined in abundance in Virginia and Alabama streams since the 1980s. Large numbers of loggerhead musk turtles were taken from Florida springs for the pet trade during the 1980s, and should that continue the populations would probably disappear.

The small size and depressed shell allow even adults to hide in rock crevices in streams.

Flattened Musk Turtle *Sternotherus depressus*

DESCRIPTION The flattened musk turtle is a small turtle with a depressed shell whose rounded edges give it a streamlined appearance. A moderate keel present down the center of the carapace in juveniles becomes less visible or absent in larger juveniles and adults. The shell is horn colored to dark brown with black streaks. The pale pink or yellowish plastron is unmarked and has one hinge. The olive-colored head may have lines, spots, or mottling. Two pairs of barbels are present under the chin.

VARIATION AND TAXONOMIC ISSUES No subspecies have been described, although the species itself was at one time considered a subspecies of the stripe-necked musk turtle. Minor variation in head patterns has been observed in different regions, despite the limited geographic range.

The pale pink or yellowish plastron is unmarked and has one indistinct hinge.

How do you identify a flattened musk turtle?

DISTINGUISHING CHARACTERS
Small, flattened brown shell

AVERAGE SIZE

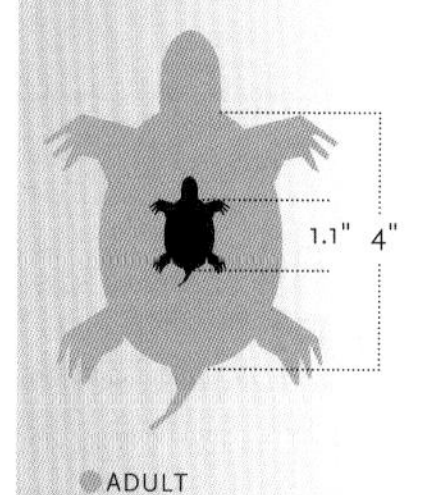

ADULT
HATCHLING

CARAPACE SHAPE

FRONT VIEW

SIDE VIEW

TOP VIEW

This turtle's unusual profile is evident.

Did you know?

Some species of turtles are confined to a single river drainage.

WHAT DO THE HATCHLINGS LOOK LIKE? Newborns resemble adults but have a more pronounced keel running lengthwise down the carapace, which flares out at the edges.

CONFUSING SPECIES Other musk turtles have carapaces that are much more elevated than the depressed shell of the flattened musk turtle. Mud turtles have two hinges on the plastron instead of one.

DISTRIBUTION AND HABITAT The flattened musk turtle has one of the most restricted ranges of any North American turtle. It occurs only in clear, shallow streams within the Black Warrior River drainage in Alabama, generally above the zone where the Cumberland Plateau meets the Gulf Coastal Plain.

BEHAVIOR AND ACTIVITY Flattened musk turtles spend much of the day hiding under sunken logs or in rock crevices underwater, a behavior for which the unusual shell shape seems well adapted. These turtles rarely bask. They seem to maintain small home ranges and return to the same hiding places repeatedly. Only nesting females are likely to venture onto land.

FOOD AND FEEDING Snails, aquatic insects, and introduced Asiatic clams seem to make up the bulk of the diet. Small native mussels were likely a primary food source before poor water quality and competition by the Asiatic clams reduced their populations.

Flattened Musk Turtle

Sternotherus depressus

REPRODUCTION Despite their small size, females take 6–8 years to reach maturity. Nests are presumably dug in sandy banks at the edge of wooded areas adjacent to the stream. Clutches containing 2 eggs have been observed. The incubation period is about 45 days, and the eggs hatch in late summer. Little else is known about nesting and reproduction in this species.

Eggs are laid in sandy soil on land near the water.

PREDATORS AND DEFENSE Nest predators probably include raccoons and other carnivorous mammals. Large fish, river otters, wading birds, and raccoons along the shoreline probably prey on young turtles and occasionally on adults.

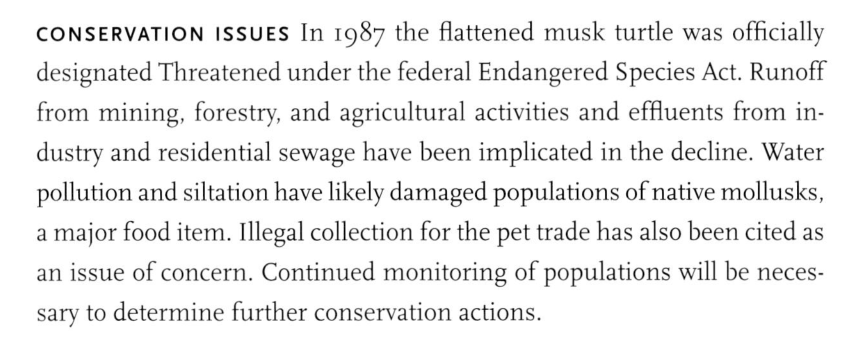

CONSERVATION ISSUES In 1987 the flattened musk turtle was officially designated Threatened under the federal Endangered Species Act. Runoff from mining, forestry, and agricultural activities and effluents from industry and residential sewage have been implicated in the decline. Water pollution and siltation have likely damaged populations of native mollusks, a major food item. Illegal collection for the pet trade has also been cited as an issue of concern. Continued monitoring of populations will be necessary to determine further conservation actions.

A spiny softshell resembles a flattened pancake.

How do you identify a spiny softshell turtle?

DISTINGUISHING CHARACTERS
Rounded, leathery shell; snorkel-like snout; spiny points on front of carapace

AVERAGE SIZE

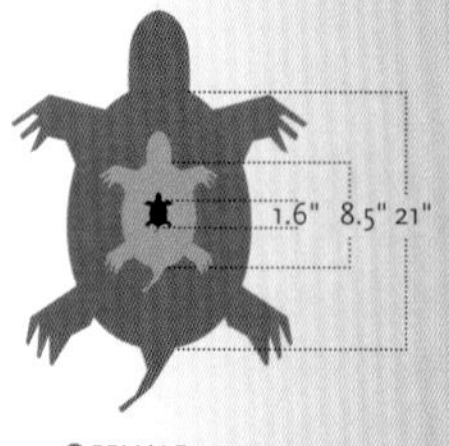

● FEMALE
● MALE
● HATCHLING

CARAPACE SHAPE

FRONT VIEW

SIDE VIEW

TOP VIEW

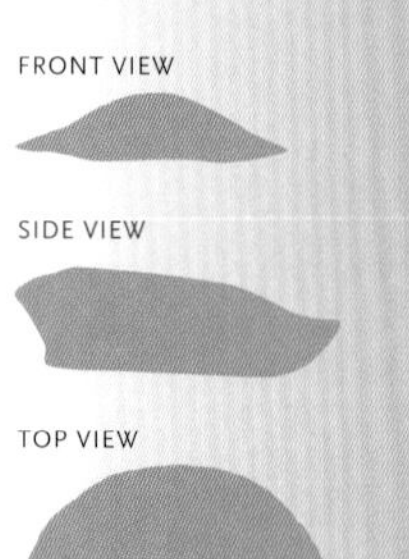

Spiny Softshell Turtle *Apalone spinifera*

DESCRIPTION Spiny softshells, like the other softshell species, have webbed feet with large claws; a flat, leathery shell ranging from brown to sand colored with a contrasting lighter-colored plastron; and a pair of light-colored stripes on the sides of the head. The limbs are gray with some dark streaks or spots. The highly flexible neck is almost as long as the body itself. The front portion of the carapace is lined with tiny spinelike protuberances and in many individuals feels as rough as sandpaper. The sex of adult spiny softshells can usually be distinguished because females are as much as one and one-half times as large as males and because males usually retain the dark circles on the carapace characteristic of juveniles, whereas females develop a mottled, camouflage-like pattern.

VARIATION AND TAXONOMIC ISSUES The spiny softshell turtle is partitioned into seven subspecies. Six of these are found in the United States, and one of them (the Gulf Coast spiny softshell, *A. s. aspera*) is found exclusively in the Southeast. The eastern spiny softshell (*A. s. spinifera*), pallid spiny softshell (*A. s. pallida*), and western spiny softshell (*A. s. hartwegi*) are variously distributed in the southeastern states. The four southeastern subspecies do not differ dramatically in appearance, the major variations being in carapace spotting and the pattern of the white or yellow head lines. The

Gulf Coast spiny softshell usually has distinctive black lines or dashes on the carapace that run parallel to the rear of the shell. The other subspecies have only a single line, but variation in these lines and dashes is not geographically consistent among the other subspecies. Most of the patterns and color variations become obscure in larger specimens of all subspecies, making differentiation difficult if not impossible.

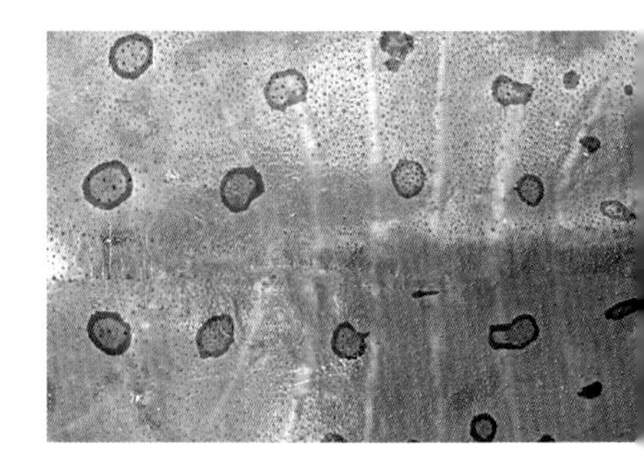

Males (top) usually retain sharply defined dark spots or circles on the carapace. Females (bottom) develop a mottled, camouflage-like pattern.

WHAT DO THE HATCHLINGS LOOK LIKE? With its round carapace and tan to yellowish color, a hatchling resembles a small pancake with legs and a head. Dark spots, some of which may be clear in the middle and appear as dark circles, are distributed over the carapace.

CONFUSING SPECIES The soft, pliable, leathery shell and snorkel-like snout of the spiny softshell immediately set it apart from the hard-shelled turtles. Spiny softshells appear round when viewed from above and have a small horizontal ridge inside each nostril; Florida softshells are more oval, and smooth softshells lack the nostril ridge as well as the spiny protrusions on the carapace.

DISTRIBUTION AND HABITAT Spiny softshell turtles have the greatest geographic range of the softshell turtles. They are found in at least portions of all the southeastern states but are absent from peninsular Florida, most of Virginia, and much of North Carolina. Spiny softshells prefer flowing water and are characteristic inhabitants of sandy-bottomed rivers that are clean and clear, but are also found in oxbows and reservoirs. They are absent from seasonal wetlands.

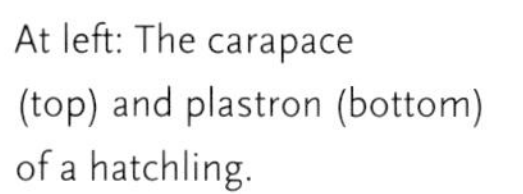

At left: The carapace (top) and plastron (bottom) of a hatchling.

BEHAVIOR AND ACTIVITY Spiny softshell turtles seldom leave the aquatic environment except to bask or lay eggs. They commonly bask on logs, sandbars, and along the shore in shallow water, but they immediately dive for deep water after spotting an observer. Adults are among the fastest swimmers of the aquatic turtles.

FOOD AND FEEDING The diet consists mostly of insects, crayfish, and fish, both large and small. Spiny softshells may capture smaller fish with a quick strike of the long neck, but larger fish are presumably scavenged. Spiny softshells can probably smell underwater by pumping water in and out through their nostrils and mouth.

In the Gulf Coast spiny softshell (right), the two head lines meet behind the eye. The lines remain parallel in the eastern subspecies (left).

A hatchling Gulf Coast spiny softshell (left) shows at least two parallel lines on the posterior of the carapace. All other subspecies have a single black line (right) along the posterior margin of the carapace.

A hatchling emerges from its hiding place on a river sandbar.

REPRODUCTION Large sandbars along the banks of large rivers are important nesting habitat. Females lay one or two clutches of 3–39 eggs in June or July. The eggs hatch in late summer or early autumn before cold weather sets in, and the hatchlings typically head straight for the water, although some may overwinter in the nest and emerge in the spring. In contrast to most southeastern turtles that have been studied, the sex of softshell turtles is not determined by the temperature of the nest.

Spines on the anterior carapace distinguish spiny softshells from other softshell turtles.

PREDATORS AND DEFENSE Typical mammals expected on a river sandbar, such as raccoons and skunks, are known nest predators. In areas where their ranges overlap, alligators are the major predators of adults. Imported fire ants, which colonize sandbars on southern rivers, are likely a problem for eggs and hatchlings. In addition to their incredible ability to completely conceal themselves in soft mud, gravel, or a sandy bottom, spiny softshells can escape predators by swimming rapidly and then biting savagely if captured.

CONSERVATION ISSUES Softshell turtles are harvested for food commercially in many regions, an activity that will require some regulation if it is to continue. Softshells are also hooked and killed by anglers. They may be more vulnerable than hard-shelled turtles to some forms of aquatic pollution because of the permeability of their leathery shell. The use of river sandbars for recreational boating and camping activities during late spring and summer could have a negative impact on nesting.

An adult female.

How do you identify a smooth softshell turtle?

DISTINGUISHING CHARACTERS
Rounded, leathery shell; snorkel-like snout; front of carapace smooth

AVERAGE SIZE

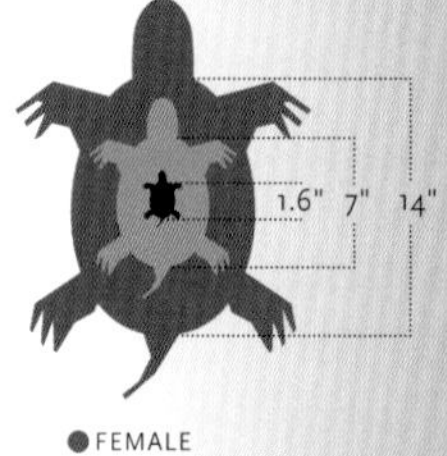

● FEMALE
● MALE
● HATCHLING

CARAPACE SHAPE

FRONT VIEW

SIDE VIEW

TOP VIEW

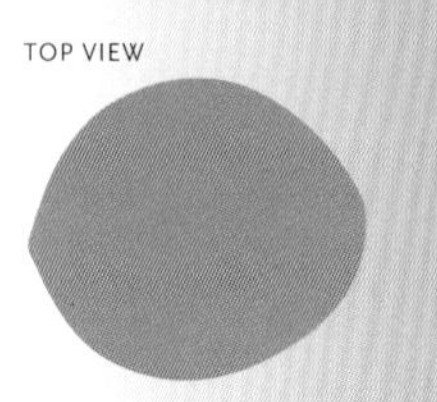

Smooth Softshell Turtle *Apalone mutica*

DESCRIPTION Adults of both sexes of the smooth softshell have round shells and are smaller than the other two species of softshells. The leathery shell, strongly webbed feet, and pointed, snorkel-like nose give them the distinctive appearance characteristic of all softshell turtles. Females often become twice the size of males. The carapace of males generally has indistinct darker markings, and a yellow stripe is present on each side of the head and neck; the carapace of females has darker mottling, and the yellow head stripes become faint or absent on the largest individuals.

VARIATION AND TAXONOMIC ISSUES The two subspecies that have been described—the Gulf Coast smooth softshell, *A. m. calvata,* and midland smooth softshell, *A. m. mutica*—differ in subtle ways that are most recognizable in the hatchlings. Young midland smooth softshells usually have a light-colored stripe on the snout between the eye and nostril and a mostly unmarked brownish carapace; hatchlings of the Gulf Coast subspecies have no stripe on the front of the snout but do have large dark dots on the carapace.

WHAT DO THE HATCHLINGS LOOK LIKE? Hatchlings look like miniature adults. Unlike other species of softshells, the underside of the leathery carapace margin is dark in contrast to the light gray or white plastron.

Smooth Softshell Turtle

Apalone mutica

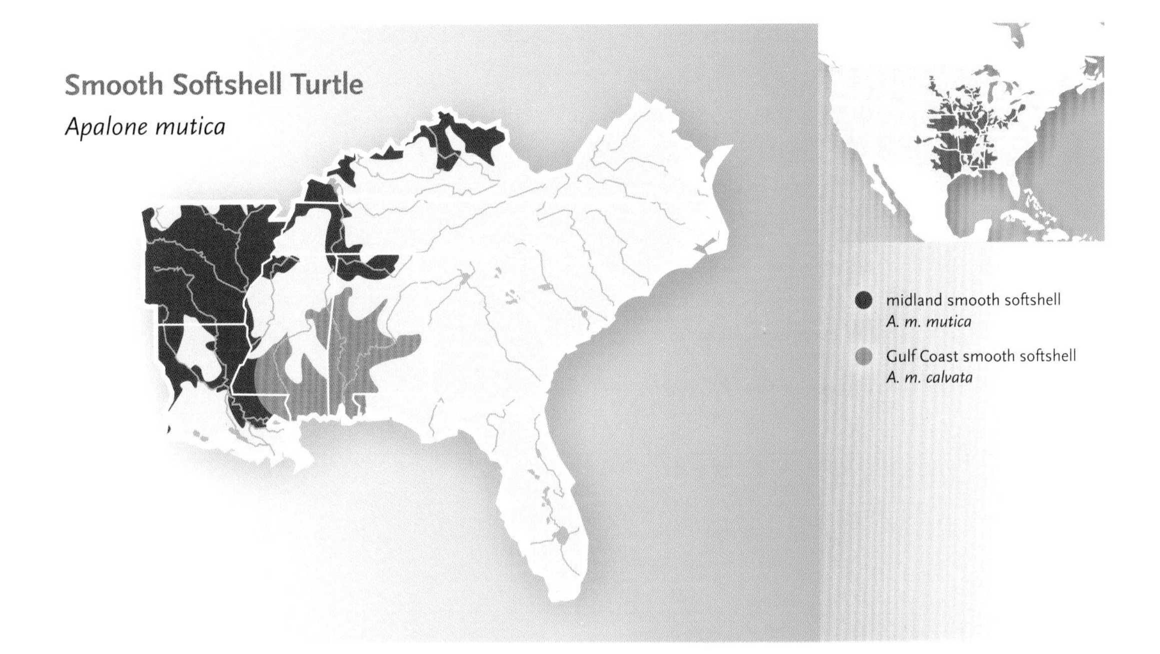

A hatchling of the Gulf Coast subspecies (top) has blotches or spots on the carapace, but the midland subspecies (bottom) has an unmarked carapace.

CONFUSING SPECIES Like the other softshell turtles, this species is readily distinguished from other North American turtles by the flattened body and leathery shell. Its completely smooth carapace without spiny projections or rounded bumps anywhere distinguishes it from the other softshells. Smooth softshells also have round nostrils with no internal ridges, and the limbs are grayish with few or no dark markings.

DISTRIBUTION AND HABITAT In the Southeast, smooth softshells are typically found in large, sandy-bottomed streams and rivers in the Ohio, Mississippi, and some Gulf Coast drainages. They are occasionally found in larger lakes but do not occur naturally in smaller ponds or seasonal wetlands.

BEHAVIOR AND ACTIVITY Smooth softshells have a reputation for speed that is unparalleled among North American turtles. Like other softshells they leave the water only to nest or bask. Individuals frequent shallow sandbars along streams and rivers where they can bury themselves underwater at a depth that allows them to stretch their neck and bring the nostrils to the surface to breathe. They often bask on sandy beaches, positioning themselves so that they can quickly retreat into the water.

FOOD AND FEEDING Smooth softshell turtles primarily eat aquatic insect larvae, which they suck up from the sand or mud bottom. They also consume worms, crayfish, berries, and seeds. Smooth softshells use their speed to pursue small fish and their long neck to strike out at them. They will also eat dead fish. Females typically forage in deeper waters while the smaller males search for prey near the shore.

A smooth softshell basks submerged on a shallow sandbar.

REPRODUCTION Smooth softshell turtles mate in spring and perhaps again in late summer. Mating occurs while a pair swims in open water; males do not grasp the anterior part of the female's shell but must continually swim with their forelimbs to maintain copulation. Females dig nests on sandbars that are above normal flood lines and lay about 15 eggs. Although the female covers the eggs after laying them, nests resemble small craters. The eggs hatch in late summer. Hatchlings open the egg from the inside using their foreclaws rather than a caruncle, as other turtle hatchlings do. The hatchlings emerge from the nest and head directly to the water under the cover of darkness.

PREDATORS AND DEFENSE Juveniles are potential prey for numerous aquatic predators such as fish, snapping turtles, and cottonmouths. Adults in large southeastern rivers have few predators, the most likely being alligators and alligator gar. Their primary defenses are concealment and speed. A threatened softshell will either conceal itself beneath the sand in shallow water or swim swiftly to deeper waters. Large individuals can deliver a powerful bite if handled.

CONSERVATION ISSUES Because they depend on clear streams and rivers, smooth softshells are vulnerable to all activities that affect the environmental health of such habitats. Pollution that affects water quality can reduce the aquatic invertebrates on which the turtles feed or may even affect softshell turtles directly. Likewise, damming river stretches where smooth softshells occur creates barriers to movement and eliminates the sandbars they need for nesting.

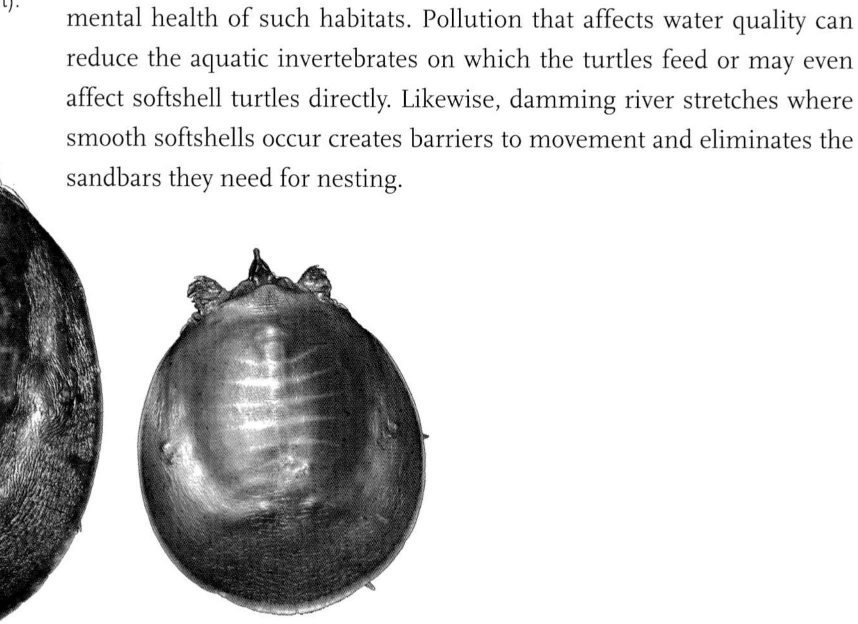

Adult females (left) are larger than adult males (right).

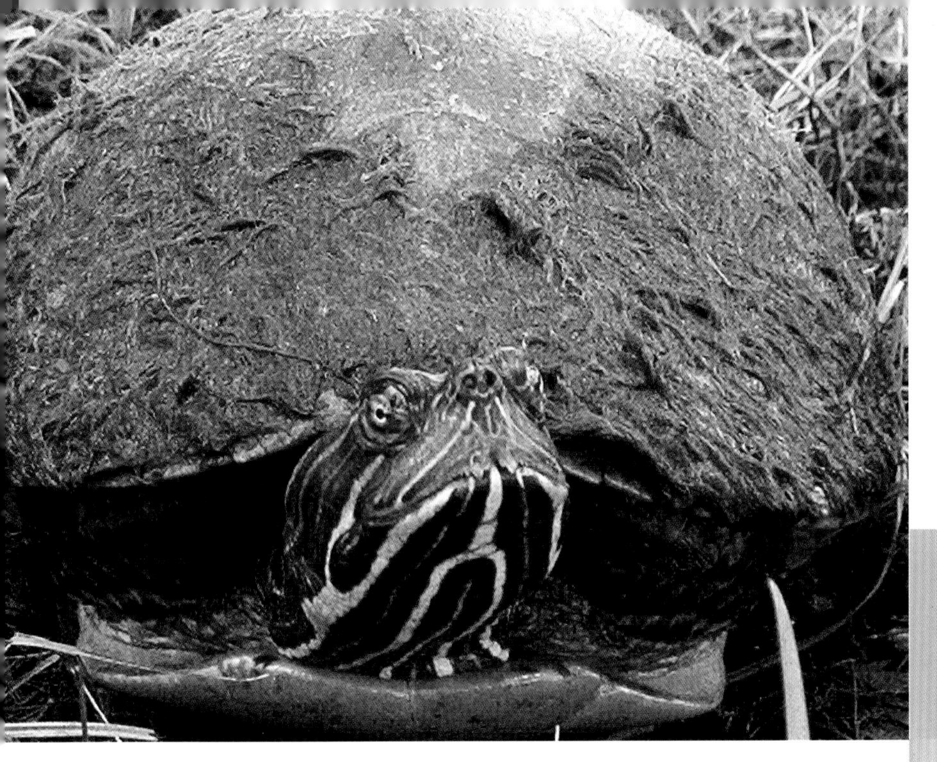

An Alabama red-bellied cooter. Note the cusps that flank the center notch on the upper jaw.

How do you identify an Alabama red-bellied cooter?

DISTINGUISHING CHARACTERS

Reddish plastron; thin yellow lines on head; notched upper jaw

AVERAGE SIZE

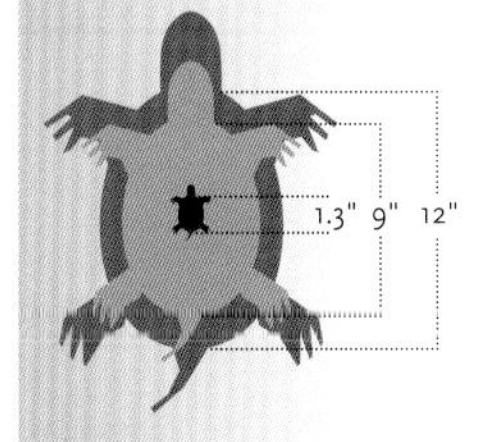

● FEMALE
● MALE
● HATCHLING

CARAPACE SHAPE

FRONT VIEW

SIDE VIEW

TOP VIEW

Alabama Red-bellied Cooter *Pseudemys alabamensis*

DESCRIPTION The Alabama red-bellied cooter is a thick-shelled turtle. The oval brown carapace develops a washboard appearance on the costal scutes, which often have orangish coloration that extends onto the marginal scutes. The plastron is reddish and may have dark markings in the seams. The head is streaked with multiple thin yellow lines, one of which is centered between the eyes and ends near the nose in an "arrow." A deep notch in the front of the upper jaw is flanked on both sides by conspicuous cusps. Females get larger than males and develop a more domed appearance. Adult males have disproportionately long, straight claws on the front feet.

VARIATION AND TAXONOMIC ISSUES No subspecies have been described, and no regional variation has been noted.

This individual's head has few lines.

Hatchlings are brightly colored with green above (top) and red below (bottom).

WHAT DO THE HATCHLINGS LOOK LIKE? Hatchlings are rounder bodied than the adults and have a brightly colored green carapace with yellow markings and a slight keel down the middle. The plastron is reddish or orangish, usually with dark markings between the scutes.

CONFUSING SPECIES The geographic range of the Alabama red-bellied cooter coincides with the ranges of several species of large, hard-shelled turtles. The carapace of adult Alabama red-bellied cooters has no elevated keel, in contrast to the shells of map turtles. The rump has yellow spots or blotches, while sliders and chicken turtles have yellow vertical stripes. River and pond cooters do not have a yellow, arrow-shaped stripe between the eyes or cusps on the upper jaw.

DISTRIBUTION AND HABITAT The Alabama red-bellied cooter is known only from the lower Mobile–Tensaw River Basin in Alabama and the lower Pascagoula River in Mississippi. The species' primary haunts are the broad freshwater channels with thick aquatic vegetation of the lower Mobile and Pascagoula river deltas, although individuals occasionally venture into brackish waters along the Gulf of Mexico.

BEHAVIOR AND ACTIVITY These turtles can be active during any month, although they become dormant along river bottoms during cold spells. Alabama red-bellied cooters bask on logs and thick vegetation during much of the year, but they are shy and will retreat into the water if approached. Some

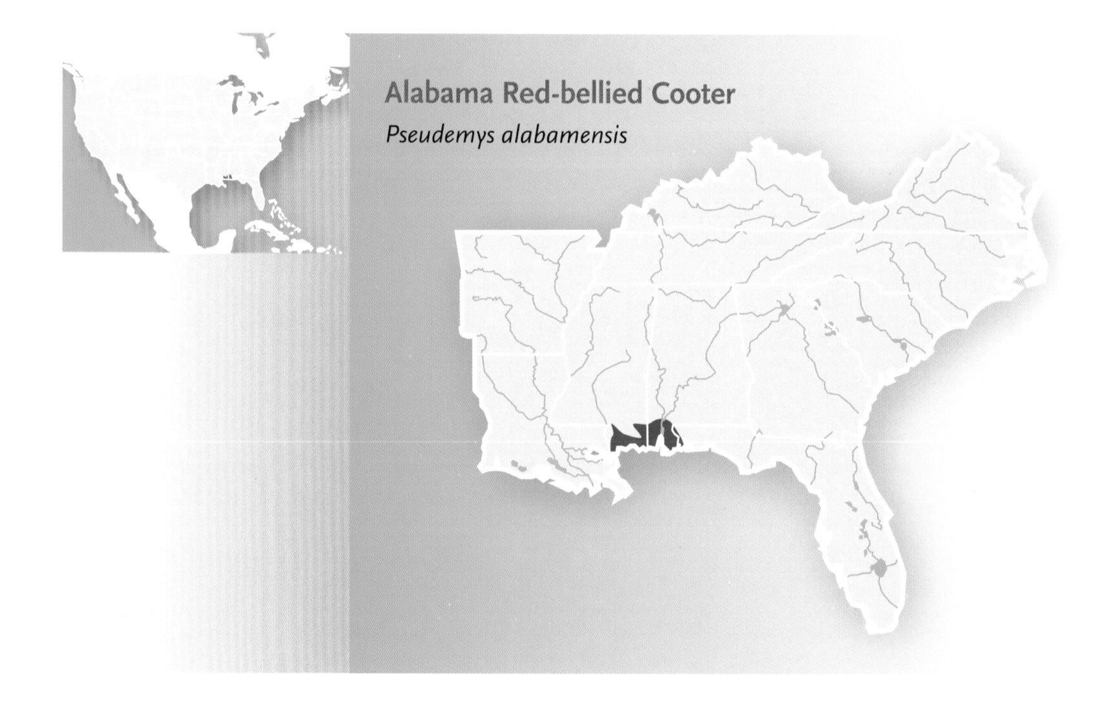

The serrated jaws are suited for shredding aquatic plants. The head on this individual is streaked with multiple thin yellow lines.

adults enter brackish waters in the estuaries, and occasional individuals are found with barnacles on their shell.

FOOD AND FEEDING Adults are typically herbivorous, feeding on submerged aquatic vegetation and algae, although they will probably eat animal prey or dead fish if these are easily obtained. Like many other species of herbivorous turtles, juveniles eat proportionately more animal matter than adults.

REPRODUCTION The reproductive cycle of the Alabama red-bellied cooter is presumed to be typical of that of other cooters. Females lay a clutch of 16–26 eggs in late spring or early summer. The best-known site where females nest in high densities is on open, sparsely vegetated beaches of Gravine Island and the Blakeley River in Baldwin County, Alabama. Females also nest on high sandbars, levees, and other elevated spots that are least likely to flood.

PREDATORS AND DEFENSE The greatest sources of mortality documented for the Alabama red-bellied cooter are fish crows and raccoons, which eat the eggs soon after they have been laid. Armadillos, introduced fire ants, feral hogs, and human disturbance may also pose a danger to eggs. Road mortality is also significant. Juveniles fall prey to large fish and cottonmouths, and even adults are prey for alligators.

CONSERVATION ISSUES The Alabama red-bellied cooter is clearly a species requiring conservation consideration. It was officially designated Endangered under the federal Endangered Species Act in 1987. Among the standard threats are fire ants, invasive plant species, pollution of the aquatic habitat, human disturbance of nesting areas, and limited availability of protected nesting sites. Solutions must be found to minimize road mortality of nesting females and emerging hatchlings on the Mobile Bay Causeway (U.S. Highway 90).

Did you know?

The Alabama red-bellied cooter and the flattened musk turtle have the smallest ranges among all North American turtles.

An adult river cooter.

How do you identify a river cooter?

DISTINGUISHING CHARACTERS
Orange plastron with black figures along seams; sleek carapace flared in back; yellow or orangish stripes on head

AVERAGE SIZE

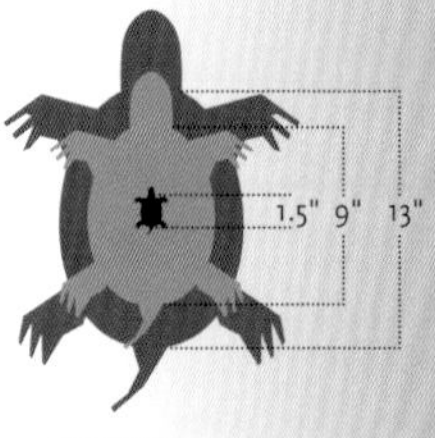

● FEMALE
● MALE
● HATCHLING

CARAPACE SHAPE

FRONT VIEW

SIDE VIEW

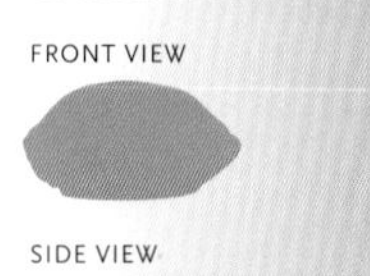

TOP VIEW

River Cooter

Pseudemys concinna

DESCRIPTION With their relatively flat shell that flares at the rear, river cooters seem streamlined for their aquatic habitat. They are highly variable in color pattern, but typical specimens have a brown to black carapace, and the plastron is yellow, orange, or reddish with prominent patterns of orange and black. Sometimes the margins of the plastron are pinkish. The head has several yellow or orangish stripes. Juveniles and young adults are more brightly colored than old adults. Hybridization with pond cooters presumably occurs throughout the Coastal Plain regions where the two species come in contact, possibly explaining why many turtles show such variation in markings and shell shapes. Females are larger on average than males, but the needlelike foreclaws of adult males are impressively long.

VARIATION AND TAXONOMIC ISSUES The regional variation, intergradation within the species, and possible hybridization among river cooters and pond cooters create a taxonomic quagmire. Some turtle biologists consider river cooters to represent at least two species; others see one species and four or five subspecies. Some consider the Suwannee River cooter, historically known as a subspecies (*P. c. suwanniensis*) on the basis of a variety of characters, to be a separate species, *P. suwanniensis*. Likewise, some view the eastern river cooter (*P. c. concinna*) and hieroglyphic river cooter (*P. c.*

hieroglyphica) as valid subspecies. The Mobile Bay cooter (*P. c. mobilensis*), which ranges from the Florida panhandle through Louisiana, is a particularly thorny taxonomic problem.

The genetic relationships of these cooters and their proper nomenclature have been further confused by the creation of large reservoirs on many southeastern rivers above the Fall Line that have allowed the pond cooter to exist outside its native range and perhaps to hybridize with the river cooter.

WHAT DO THE HATCHLINGS LOOK LIKE? The shell is mottled green and gray-black; the pattern of yellow to yellow-orange lines on head and limbs is similar to that of the adults. The plastron of some hatchlings is orange with a gray-black center figure.

CONFUSING SPECIES River cooters can usually be distinguished from pond cooters because they do not have the latter's immaculate yellow plastron. The underside of the river cooter's marginal scutes is yellow to yellow-orange with dark marks that have light-colored centers; this character distinguishes it from all other hard-shelled turtles within its geographic range. River cooters often have concentric yellow lines on the carapace scutes, while pond cooters usually have straight lines.

DISTRIBUTION AND HABITAT In the Atlantic Coast states, from Virginia to Georgia, the river cooter is found in large rivers, generally above the Fall Line. An isolated population inhabits the New River of Virginia. River cooters prefer clear flowing water with gravel and cobble bottoms and beds of rooted aquatic plants, such as elodea and eelgrass. They are often abundant where rock ledges extend across river channels where the Piedmont meets the Coastal Plain. However, their range also extends onto the Coastal Plain, especially in the Gulf Coast region from Alabama to Louisiana. They occur in the Ouachita and Ozark mountain drainages in Arkansas and are found in portions of the Cumberland Plateau in Kentucky and Tennessee. In western Florida, the Suwannee River cooter is found in blackwater rivers that flow into the Gulf of Mexico.

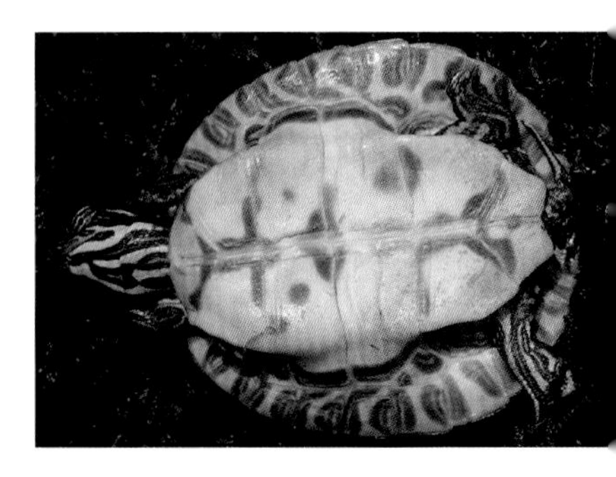

This juvenile river cooter (top) from the Piedmont of Alabama has a brilliant red pattern on its carapace. A hatchling Suwannee River Cooter (middle). The plastron pattern (bottom) is usually more distinct in juveniles than adults.

A backward "C" (top) on the carapace scutes is often used to help distinguish river cooters from pond cooters and red-bellied turtles.

The head markings on river cooters (bottom) often consist of thick dark yellow-orange lines.

BEHAVIOR AND ACTIVITY The river cooter is the species people see basking in large numbers on rock outcrops and ledges in big rivers as they cross interstate highway bridges along the Fall Line. Notable turtle concentrations can be seen on the Savannah River from Interstate 20 and on the Chattahoochee River from Interstate 85. This very wary turtle will dive from rocks when people and canoes are still hundreds of yards away. River cooters seldom go on land except to nest. Suwannee River cooters have long been known to enter brackish waters, and individuals have been found with barnacles on their shells, suggesting a sustained period in salt water.

Rock ledges at the Fall Line of all major rivers in the Southeast will have river cooters basking on them. These cooters were photographed on the Savannah River.

FOOD AND FEEDING Hatchlings and juveniles are omnivorous, but adults are primarily herbivorous. River cooters in the Savannah River bite onto blades of eelgrass and shred them as the current pulls them downstream. Suwannee River cooters have been observed feeding on turtle grass in salt water at the mouths of Florida rivers.

REPRODUCTION River cooters presumably mate in the water in spring. Females nest in open, sparsely vegetated areas such as

A Suwannee River cooter basking.

River Cooter

Pseudemys concinna

riverine sandbars, levees, and agricultural fields in floodplain areas that are above the flood zone. They generally begin nesting during the day but may not complete the process until after dark. A typical clutch has about 20 eggs, and females may lay one or two clutches each year. The eggs hatch in the fall, and some hatchlings remain in the nest until the following spring. Suwannee River cooter females often dig an unusual nest with one main chamber bordered by one or two satellite nests a few inches to each side. Most of the eggs are laid in the main nest, but 1–3 eggs may be deposited into each satellite hole. The biological significance of this nesting behavior is unknown.

River cooters usually have some black plastral markings and varying degrees of orange tinge. A river cooter (left) and a Suwannee River cooter (right).

PREDATORS AND DEFENSE Eggs in the nest are vulnerable to raccoons, skunks, and opossums. Small juveniles are eaten by carnivorous mammals and fish, wading birds, and a variety of snakes. Adult cooters are prey for large alligators and probably rely more on speed than on their shell to protect them.

CONSERVATION ISSUES As a predominantly river-dwelling species, the river cooter is subject to unnatural flooding regimes due to upstream dams, aquatic pollution, and shooting of basking turtles for sport. The Suwannee River cooter once occurred in large numbers in the Suwannee River system, but the population has been reduced by harvest for food.

Map Turtles and Southeastern Rivers

All map turtles are members of the genus *Graptemys*. The genus is nearly endemic to the United States, as only 1 of the 13 species—the common map turtle—extends its range into southern Canada. Eleven species occur in the Southeast, and the other 2 are found in eastern Texas. Map turtles are primarily river species and rarely travel overland. In the Southeast, most species are restricted to a single major river system from the Chattahoochee on the Georgia-Alabama border to the Sabine on the Louisiana-Texas border. All of these systems drain into the Gulf of Mexico, and each contains a different map turtle species, or sometimes two. The impressive biodiversity of map turtles is one reason the Southeast has world standing among areas of great turtle diversity.

The Barbour's map turtle is one of 11 species of map turtles found in the Southeast.

When Fred Cagle of Tulane University described the Ouachita and Sabine map turtles in 1953, more than half the species of map turtles recognized in 1995 had not yet been described. The taxonomy remains unsettled, but new morphological and genetic studies may help to resolve it.

Most map turtles can be partitioned into big-headed and narrow-headed groups. Females of Barbour's, Alabama, Gibbons', and Ernst's map turtles have enormously wide heads with thick jaws and crushing surfaces, and specialize on large freshwater mussels found on the river bottom. The tiny males, with normal-sized heads, eat aquatic insects such as mayflies, stoneflies, and caddisflies that they pick from log snags and rocks. The narrow-headed species include the black-knobbed, ringed, and yellow-blotched map turtles. This group is also characterized by elevated dorsal scutes, which often appear as knobs or perhaps fins; collectively they are called the "sawbacks." The remaining species, including the common, false, and Ouachita map turtles, fall into an intermediate category in which the females' heads are slightly enlarged and the males are nearer in body size to females. These intermediate-sized species tend to have the most generalist diets, eating anything from freshwater mussels to crayfish to insects to algae. They also have the broadest geographical ranges of the map turtles.

Most map turtles, especially the big-headed species, consume freshwater mussels, snails, and crayfish. The diversity of map turtles in the southeastern United States is at least partly a result of the diverse mollusk fauna originally found here. Chronic pollution and sedimentation have caused long-term declines in freshwater mollusks, which are today recognized as among the world's most endangered animals. Ironically, the colonization of many U.S. rivers by the Asiatic clam, *Corbicula fluminea*, may have an unintended but timely benefit for map turtles.

Because map turtles depend on river ecosystems, they are highly susceptible to human impacts and may function as bioindicators. Pollution from sewage, agriculture, industrial chemicals, and road runoff has degraded rivers and reduced the biodiversity of native plants and animals. The construction of dams has altered flow patterns, generally increasing summer flows—and drowning turtle nests. Sandbars, which are critical nesting areas, have been degraded by invasive vegetation, fire ants, and frequent human recreational use. Map turtles are often used for target practice and shot from their basking logs. These impacts have combined to cause declines in the Southeast's map turtle species.

Opportunities for restoration of the rivers of the Southeast are abundant. Education and regulation may be part of the solution, but recognition that southeastern rivers are valuable resources for a vast and unusual array of native wildlife such as map turtles must come from the hearts of the southerners who live along them.

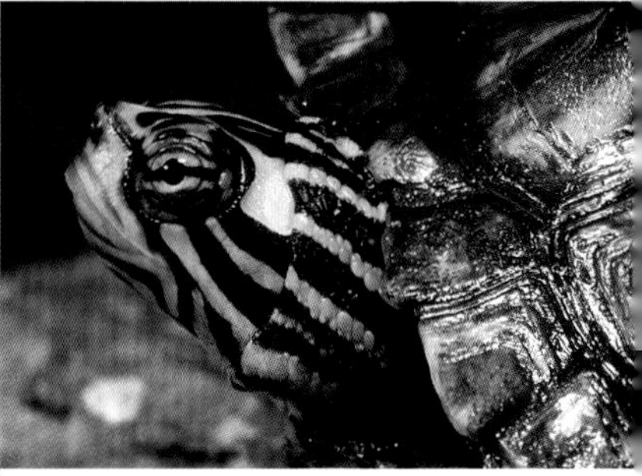

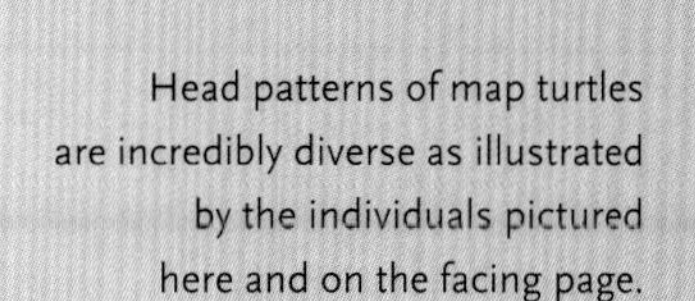

Head patterns of map turtles are incredibly diverse as illustrated by the individuals pictured here and on the facing page.

A young male basks on a log.

How do you identify a Barbour's map turtle?

DISTINGUISHING CHARACTERS
Females with massive head; greenish brown keeled carapace; greenish blotch behind each eye

AVERAGE SIZE

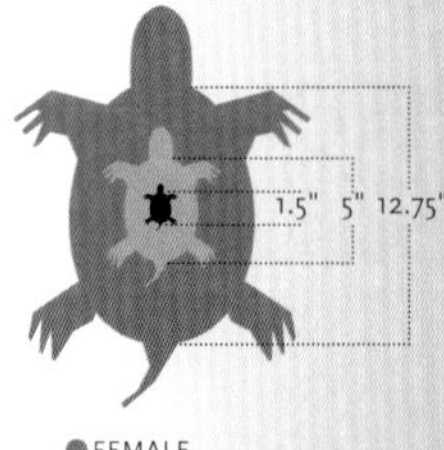

● FEMALE
● MALE
● HATCHLING

CARAPACE SHAPE

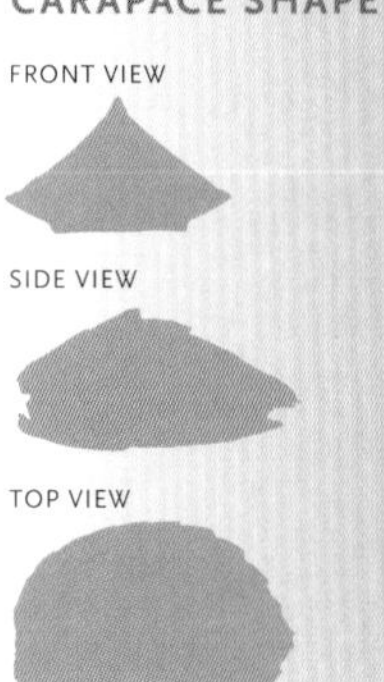

Barbour's Map Turtle *Graptemys barbouri*

DESCRIPTION Barbour's map turtle displays the greatest degree of sexual size dimorphism of any North American turtle. Females have a grossly enlarged head and prominent broad jaws, and are up to four times larger than males in body size. Males seem tiny relative to the immense females and lack the wide head. They do not have elongated front claws. A large blotch or crescent behind each eye is usually green or chartreuse, although in some it is occasionally yellow or even orange. A narrow interorbital blotch on top of the head ends in a point above the nose. A yellow transverse bar is apparent under the chin. The carapace is keeled and usually brown, and the plastron is yellow. Both sexes have spines down the center of the carapace. Barbour's map turtles belong to the group of broad-headed map turtles that includes Alabama, Ernst's, and Gibbons' map turtles. All exhibit extreme sexual size

A big-headed female basks on a limestone outcrop along a river in southwest Georgia.

dimorphism and a head pattern that includes large blotches both behind and between the eyes.

VARIATION AND TAXONOMIC ISSUES No subspecies are recognized.

WHAT DO THE HATCHLINGS LOOK LIKE? Hatchlings resemble the adults, but both sexes have normal-sized heads. A dorsal keel with very prominent black-tipped projections is often evident in hatchlings but diminishes with size and age.

CONFUSING SPECIES River cooters and yellow-bellied sliders are similar in body size, but their shells are not strongly keeled and the females do not have enlarged heads. In fact, no other hard-shelled turtle in the range of Barbour's map turtle has an enormously enlarged head except in the Choctawhatchee River system, where it co-occurs and may hybridize with Ernst's map turtle.

Vertebral spines are evident on hatchlings (left). The plastron of a hatchling (right).

The female's exceptionally large head is effective for crushing mussels and snails.

The tiny males eat aquatic insects such as caddisflies and dragonfly nymphs.

DISTRIBUTION AND HABITAT Barbour's map turtles occur in the Gulf Coastal Plain in the Apalachicola River system, including the Chattahoochee, Flint, and Chipola Rivers in eastern Alabama, western Georgia, and western Florida. They have recently been reported from the Choctawhatchee River system of southeastern Alabama and the adjacent Florida panhandle. These rivers are clear flowing and have limestone rock and cobble bottoms and many fallen trees.

BEHAVIOR AND ACTIVITY Both sexes bask frequently. Females often select isolated rocks or tree stumps far from shore; males and juveniles are more likely to be found among tree trunks with a network of smaller branches. Barbour's map turtles are wary and will dive from basking sites at the slightest hint of danger, often dropping straight to the bottom and hiding among rocks and logs. In comparison, river cooters, which frequently bask with them, will drop from the basking site and swim for a hundred feet or more and hide in the deepest pools.

FOOD AND FEEDING The characteristic wide head of the females provides attachment for massive jaw muscles and crushing plates, allowing them to specialize on a diet of aquatic snails and freshwater mussels. The crushed shell fragments are swallowed and passed through the digestive system, and piles of turtle scats containing them are a sure sign of favorite feeding and basking areas. The females also scrape off and eat freshwater sponges attached to rocks and logs. The tiny males eat aquatic insects such as caddisflies and dragonfly nymphs that they snatch from limestone rock ledges, rocks in riffles, and sunken logs.

Females dig their nests on sandbars and beaches.

REPRODUCTION Barbour's map turtles nest during June–August. The females dig their nests on the tall sandbars and beaches that line river courses and lay multiple clutches of 4–11 eggs. The eggs hatch in late August and September. The small males may mature at 4 years of age; females may take 15–20 years to reach maturity.

PREDATORS AND DEFENSE Raccoons and other mammalian predators will destroy nests, as do introduced fire ants. Adults disturbed while basking tend to dive and hide among rocks on the river bottom. Females often display a formidable open mouth when handled; males retreat passively into the shell.

Did you know?

Some turtles live as long as many humans and do not reach maturity until they are teenagers.

CONSERVATION ISSUES Although small male Barbour's map turtles may find abundant aquatic insects within the man-made reservoirs that have replaced stretches of the shallow, free-flowing rivers in which these turtles live, deep reservoirs with silt bottoms cannot support the freshwater mollusks the females require. Thus, females cannot survive in reservoirs, and populations of Barbour's map turtles decline in those impounded regions. River pollution has likely had an impact on this species, and examples of unexplained shell rot that may have been caused by it have been noted. Collection for the pet trade may have reduced the size of some local populations. In recent years, invasive plants have colonized many river sandbars, especially those no longer subject to natural flooding. These plants make nest construction difficult and shade potential nest sites. Fire ants may exact an unknown toll on eggs and hatchlings. Clean, naturally flowing river sections with freshwater mollusks are essential for the long-term persistence of Barbour's map turtles.

The plastron of a two-year-old male.

A female heads back to the water after nesting on a sandbar.

How do you identify an Alabama map turtle?

DISTINGUISHING CHARACTERS
Yellow marks on each marginal scute; drab brown to olive carapace; greenish blotch on head; females have large head

AVERAGE SIZE

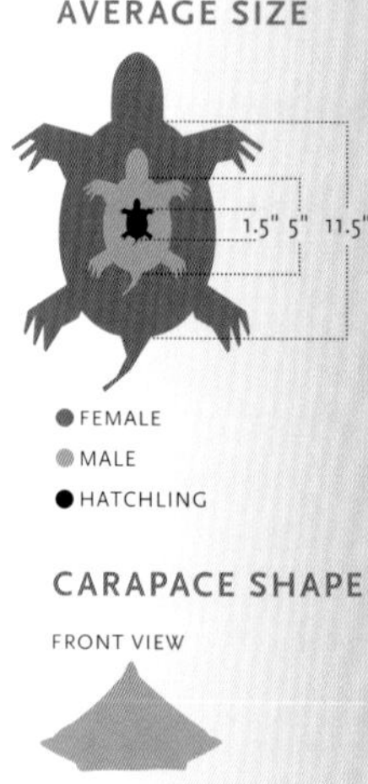

CARAPACE SHAPE

FRONT VIEW

SIDE VIEW

TOP VIEW

Alabama Map Turtle *Graptemys pulchra*

DESCRIPTION Alabama map turtles are large hard-shelled turtles that are strongly sexually dimorphic. Females have a large head and prominent broad jaws and are typically two or more times larger than the males, which do not have enlarged heads. Both sexes have a pointed snout. The front of the drab brown to olive carapace has a moderate-sized hump. The plastron is yellow with a black pattern emanating from the seams of the scutes. A large, continuous greenish blotch encompasses the sides and top of the head. A small black "Y" on top of the head extends from the neck toward the eyes but does not contact the orbits. Each marginal scute has a pattern of concentric yellow circles. The spines down the center of the carapace are present in both sexes. Males do not have elongated front claws.

VARIATION AND TAXONOMIC ISSUES No subspecies have been described, and no substantive geographic variation is known.

WHAT DO THE HATCHLINGS LOOK LIKE? Hatchlings resemble the adults but are more strikingly colored.

A hatchling (top) with a distinctly raised spine.

This recently hatched Alabama map turtle (bottom) is still absorbing its yolk.

Alabama Map Turtle

Graptemys pulchra

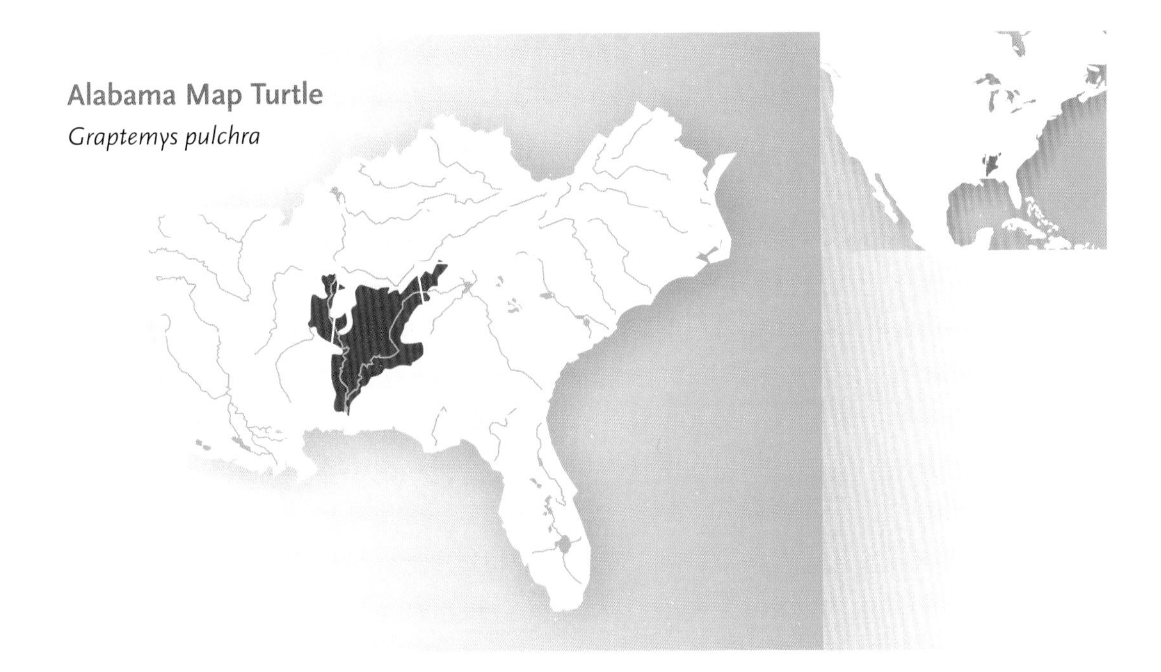

CONFUSING SPECIES Alabama red-bellied turtles and river cooters are similar in body size, but no other hard-shelled turtle in the range of the Alabama map turtle has an enormously enlarged head. In the upper drainages of some Alabama rivers where they co-occur, common map turtles have moderately large heads. Black-knobbed map turtles, which are smaller, have yellow stripes on the sides of the head rather than broad greenish blotches.

A male Alabama map turtle.

DISTRIBUTION AND HABITAT Alabama map turtles occur in the Alabama, Cahaba, Tombigbee, Coosa, lower Tallapoosa, and Black Warrior river systems. All of these river systems are on the Gulf Coastal Plain or Cumberland Plateau and drain into Mobile Bay. Flowing water and swift currents are preferred, although these turtles do occur in reservoir sections.

BEHAVIOR AND ACTIVITY Alabama map turtles bask on logs and brush where they can quickly escape into fast-moving water. They are active from March to November, but little else is known about their ecology and behavior.

Alabama map turtles have a pointed snout.

A large female with characteristic head pattern.

FOOD AND FEEDING The large-headed females likely specialize on native mussels and snails. Smaller males and juveniles probably eat aquatic insects.

REPRODUCTION Nesting begins in late April and continues into July or August. Sandbars and sandbanks above flood lines are likely nesting habitats. Females may lay multiple clutches of about 5 eggs each. Very little is known about the reproductive ecology of this species.

PREDATORS AND DEFENSE Raccoons, armadillos, common crows, and fish crows are known nest predators.

CONSERVATION ISSUES The Alabama map turtle is affected by the degradation of rivers and streams that has led to declines in freshwater mussels. The loss of sandbar nesting habitat, through dams and the alteration of the natural flooding regime they cause, undoubtedly affects these turtles. Removal of log snags for navigation and recreation eliminates structure needed for basking and habitat structure for the aquatic insects the males eat. Use of map turtles for target practice is known to occur.

An adult female.

Ernst's Map Turtle

Graptemys ernsti

DESCRIPTION Ernst's map turtles have a plain yellow plastron, usually without any pattern. The relatively high-domed carapace is drab brown to olive with some vertical yellow bars on the scutes on either side and a vertical yellow bar near the front edge of each marginal scute. Greenish blotches behind each eye are separated from a blotch between the eyes on the top of the head by a V-shaped black bar that contacts the top of each eye. The blotch on the top of the head has three prongs that extend to the edge of the snout. A yellow horizontal bar extends back from the eye on each cheek. The snout appears pointed, and both sexes have spines along the center of the carapace. Ernst's map turtles are strongly sexually dimorphic. Females have a large head; broad, wide jaws; and are two or more times larger than the males, which do not have an enlarged head. Males do not have elongated front claws. Ernst's map turtle is also called the Escambia map turtle by some authorities.

The pale yellow plastron of Ernst's map turtles can have some black pigment along the seams.

How do you identify an Ernst's map turtle?

DISTINGUISHING CHARACTERS
Similar to Alabama map turtle but found only in Escambia River drainages

AVERAGE SIZE

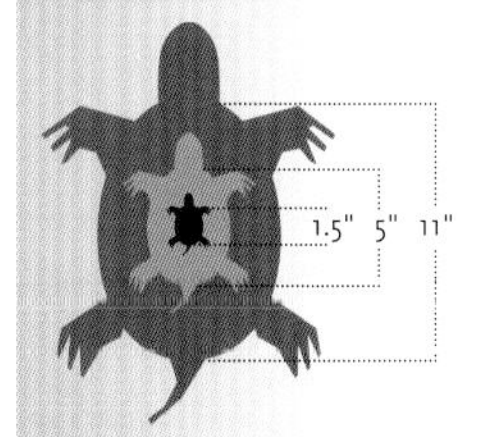

● FEMALE
● MALE
● HATCHLING

CARAPACE SHAPE

FRONT VIEW

SIDE VIEW

TOP VIEW

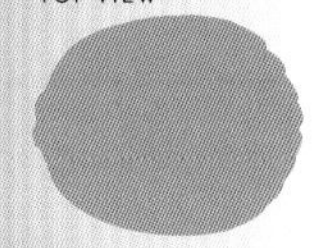

A yearling.

VARIATION AND TAXONOMIC ISSUES No subspecies are recognized. Ernst's map turtle is similar to but recognized as distinct from the Alabama map turtle and Gibbons' map turtle, the species it most resembles. These species do not co-occur in the same drainages.

WHAT DO THE HATCHLINGS LOOK LIKE? Hatchlings resemble the adults but have higher, more distinctive keels on the carapace.

CONFUSING SPECIES River cooters and yellow-bellied sliders are similar in body size but do not have enlarged heads. No other hard-shelled turtle in the range of Ernst's map turtle has an enormously enlarged head, except in the Choctawatchee and Pea rivers of Alabama, where Barbour's map turtle has recently been found and hybridization apparently is occurring. Males and smaller females have a pronounced carapace keel on some vertebral scutes that cooters and sliders lack.

This male displays typical head blotches and bars.

DISTRIBUTION AND HABITAT Ernst's map turtles are found on the Gulf Coastal Plain in the Escambia, Yellow, Conecuh, Choctawatchee, and Pea rivers of the Florida panhandle and adjacent Alabama. River habitats in which these turtles occur are large, relatively fast flowing, and have sand or gravel bottoms with abundant log snags. Hatchlings prefer sluggish waters with silt bottoms.

The unique head blotch shape helps to identify the Ernst's map turtle. Females (as shown here) have larger heads than males.

Ernst's Map Turtle

Graptemys ernsti

Ernst's map turtles are most conspicuously active during the day while basking.

BEHAVIOR AND ACTIVITY Ernst's map turtles are most conspicuously active during the day while basking, but they may feed primarily at night. They bask regularly on log snags in open water, even during the winter on warm days. They are wary and dive into the water with little provocation, although males may be less willing to desert their basking spots. Both all-male and all-female aggregations in the water have been observed, but the significance of this behavior is unknown.

FOOD AND FEEDING As might be expected from their large jaws, adult females eat native freshwater mussels, crayfish, and introduced Asiatic clams. Males and young juvenile females tend to feed on aquatic insects such as dragonfly nymphs, beetles, caddisflies, and small snails.

REPRODUCTION Little is known about the reproductive behavior of Ernst's map turtles in the wild. Courtship, in which the male vibrates his head against the female's snout, has been observed in the fall but likely can occur any time of the year the turtles are active. Females have been observed nesting during late April–July, although they may make earlier forays onto land to search for suitable nesting sites. Nests, which are typically placed in fine sand on large sandbars 6–15 feet above the water line, can survive flooding for up to a week. Females lay multiple clutches of 6–13 eggs. Hatchlings have been observed in the wild in late September and presumably do not overwinter in the nest.

A juvenile.

PREDATORS AND DEFENSE Raccoons, fish crows, common crows, and armadillos are the major nest predators.

CONSERVATION ISSUES Ernst's map turtle has a limited range. Thus, pollution, impounding of free-flowing streams, and loss of native foods could endanger it. Monitoring will be necessary to assess the trends and status of populations.

The relatively high-domed carapace has black spines along the midline.

Gibbons' Map Turtle *Graptemys gibbonsi*

DESCRIPTION Gibbons' map turtle is similar to but recognized as distinct from the Alabama map turtle and Ernst's map turtle. The relatively high-domed carapace is olive, brown, or green and has black spines along the midline. Each marginal scute has a vertical broad yellow or orange bar near the middle. The plastron is yellow, but with black markings at the seams. Large greenish blotches behind each eye are connected on top of the head. A three-pronged pattern extends toward the pointed nose but does not reach it. Gibbons' map turtles are strongly sexually dimorphic. Females have prominent broad, wide jaws; a large head; and are two or more times larger than the males. Males do not have an enlarged head or elongated foreclaws. Some authorities call this species the Pascagoula map turtle.

Large greenish blotches behind each eye are connected on top of the head.

How do you identify a Gibbons' map turtle?

DISTINGUISHING CHARACTERS
Similar to Alabama map turtle but found only in Pascagoula and Pearl river drainages

AVERAGE SIZE

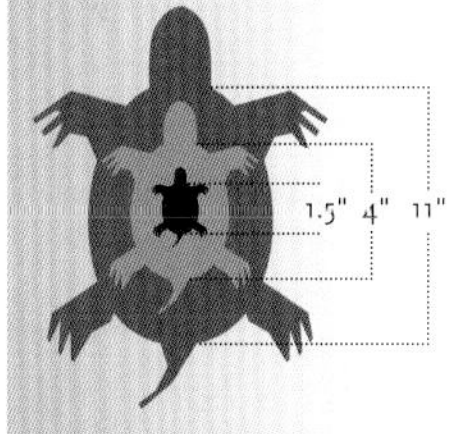

● FEMALE
● MALE
● HATCHLING

CARAPACE SHAPE

FRONT VIEW

SIDE VIEW

TOP VIEW

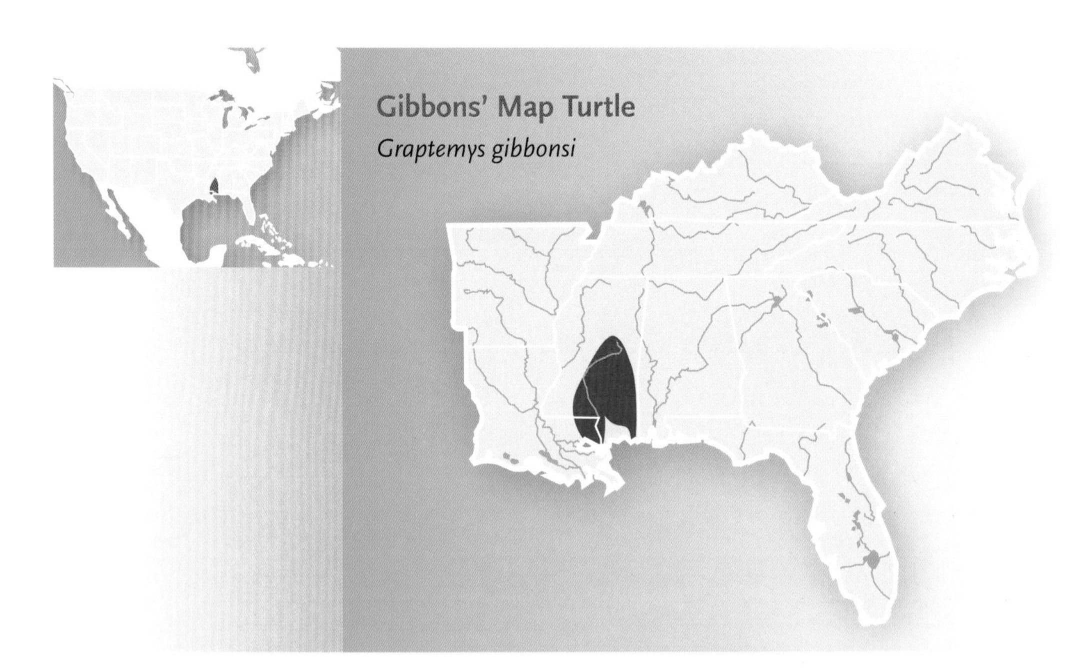

A hatchling from the Pearl River: carapace (right), plastron (left), and head and carapace (middle).

VARIATION AND TAXONOMIC ISSUES No subspecies are recognized. Gibbons' map turtle may be more similar to Ernst's map turtle than to the Alabama map turtle, even though the range of the Alabama map turtle lies between the ranges of the other two.

WHAT DO THE HATCHLINGS LOOK LIKE? The hatchlings are similar to the adults, although each carapace costal scute may have curved or reticulate yellow-orange lines. These markings tend to fade with age.

CONFUSING SPECIES Adult river cooters and red-eared sliders are about the same size as adult female Gibbons' map turtles but do not have an enlarged head. Ringed and yellow-blotched map turtles are smaller and have distinct knobs or fins on the spine of the carapace, yellow head stripes, and a small yellow blotch on the sides of the head. Gibbons' map turtles have a broad yellow to greenish blotch on each side of the head.

DISTRIBUTION AND HABITAT Gibbons' map turtles were originally described from the Pascagoula River but were later discovered in the Leaf

The plastron of a Gibbons' map turtle is yellow with black markings at the seams.

and Chickasawhay rivers in Mississippi and the Pearl and Bogue Chitto rivers of Mississippi and Louisiana. The species' range lies within the Gulf Coastal Plain. These turtles prefer deep, flowing waters with sand and gravel bottoms, and access to basking logs. They rarely if ever visit backwaters or floodplain swamps.

BEHAVIOR AND ACTIVITY Gibbons' map turtles bask frequently. They rest at night on the underwater portions of favorite basking logs and tree limb debris.

The head pattern of a Gibbons' map turtle.

FOOD AND FEEDING Males feed on aquatic insects; females specialize on large mussels and snails.

REPRODUCTION Little is known about the reproductive and nesting ecology of Gibbons' map turtle. Presumably, females nest on sandbars above the flood line.

PREDATORS AND DEFENSE Raccoons and fish crows are likely predators of eggs, and raccoons, river otters, and wading birds probably eat hatchlings. Alligators may occasionally capture an adult.

CONSERVATION ISSUES Like other large-headed map turtles, Gibbons' map turtle populations have suffered where water pollution has led to declines in native mollusks and aquatic insects associated with rivers. The lack of snags in dredged rivers may limit basking opportunities, in turn affecting body temperature and reproduction. Interestingly, museum specimens from the early 1900s are larger than turtles measured in the early 2000s, suggesting that declining habitat quality and food resources are producing smaller animals. Poor conditions might result in a shortened life span so that the turtles do not attain the maximum body size possible. Nesting beaches within the range of this turtle are degraded by campers and boaters, invasive plants, and fire ants. This species is likely in need of conservation action because the habitats where it occurs are in need of restoration.

A common map turtle from Arkansas.

How do you identify a common map turtle?

DISTINGUISHING CHARACTERS
Olive carapace with midline keel; maplike pattern on carapace

AVERAGE SIZE

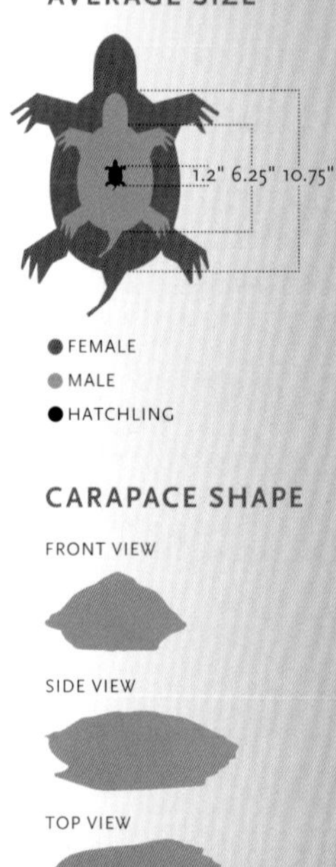

● FEMALE
● MALE
● HATCHLING

CARAPACE SHAPE

FRONT VIEW

SIDE VIEW

TOP VIEW

Common Map Turtle *Graptemys geographica*

DESCRIPTION Common map turtles are the species that gave the map turtles their name. The olive green carapace has a shallow but unmistakable keel along the midline and an intricate series of faint yellow lines that resemble a topographic map. The plastron is pale yellow, and the olive to dark brown skin is punctuated with yellow or greenish stripes. The small postorbital blotch on the head is distinctive and consistent, resembling a triangle with rounded points (one downward, one forward, one rearward) that is longer from front to back than it is high. Also very consistent is the yellow line on the lower side of the neck that arches upward as it nears the head, forming a J. The jaws are moderately large with thick surfaces for crushing snails and crayfish. Females are larger than males, but the size difference is not as pronounced as in other map turtle species.

The head of the female common map turtle is moderately large, but proportionately smaller than that of the female Barbour's map turtle.

Common Map Turtle
Graptemys geographica

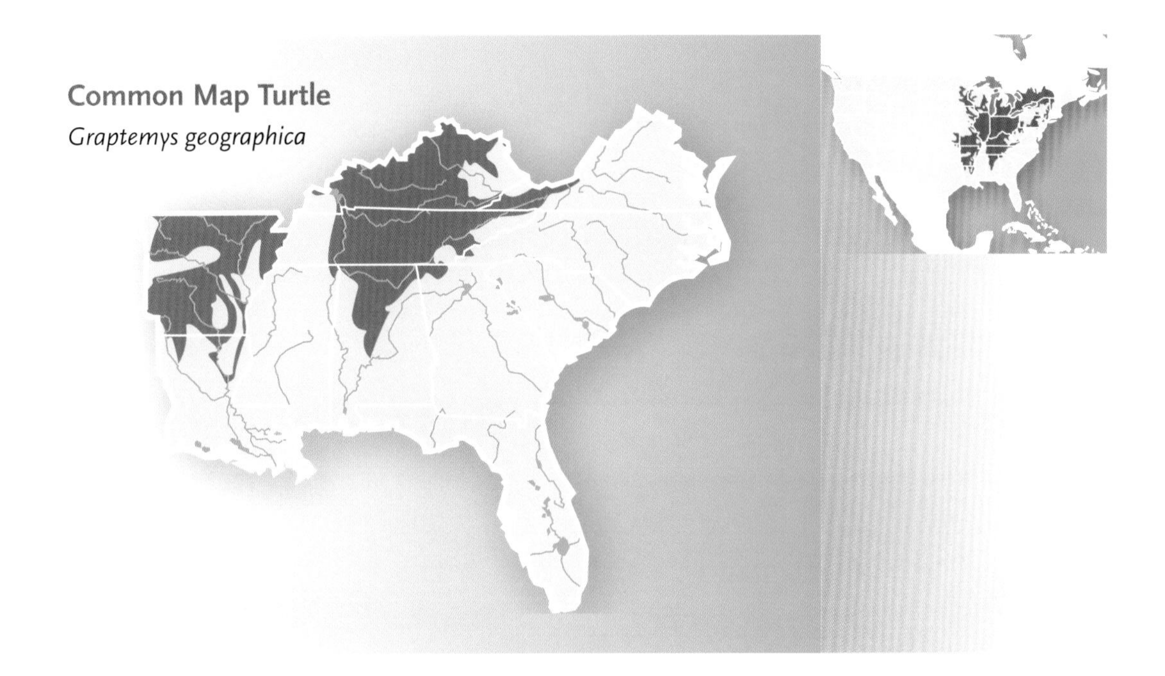

VARIATION AND TAXONOMIC ISSUES No subspecies are currently recognized, and no regional variation is detectable within the Southeast.

WHAT DO THE HATCHLINGS LOOK LIKE? Hatchlings closely resemble the adults but are perhaps more strikingly green. The seams on the plastron are prominently outlined in black, but those markings fade with age.

CONFUSING SPECIES In the Southeast, common map turtles may co-occur with false map turtles and Ouachita map turtles in parts of Arkansas, Tennessee, and Kentucky, and with Alabama map turtles in northern portions of the Mobile Bay drainages. Although similar in size and appearance, false map turtles have a more prominent vertebral keel on the carapace and a hooklike yellow patch or partial crescent behind the eye rather than a series of fine lines and a small spot. The Ouachita map turtle has a large yellow blotch behind each eye.

DISTRIBUTION AND HABITAT The common map turtle is the most widespread of all the map turtles. It is found in rivers, streams, and lakes as far north as Quebec and as far south as Alabama and Arkansas, and is the only map turtle that inhabits drainages that empty into the Atlantic Ocean, although those drainages are found in the northeastern United States. In the Southeast, it occurs in several drainages that empty into the Gulf of Mexico but is absent from much of the Mississippi River and its immedi-

The plastral pattern (top) of a juvenile.

A yearling (bottom).

A juvenile female.

A male is identified by the thick tail.

ately adjacent drainages. Common map turtles prefer clear, flowing water and gravel substrates.

BEHAVIOR AND ACTIVITY In clear, shallow, cobble-bottomed streams, juveniles often hide under big, flat rocks. Common map turtles frequently bask and are easy to see by scanning log snags along the shoreline with binoculars. During winter, they hibernate exposed on river bottoms, absorbing enough oxygen through their skin to fuel their reduced metabolism. Unlike painted turtles and common snapping turtles, they do not bury themselves in the mud during hibernation.

FOOD AND FEEDING Snails appear to be the favorite food, although freshwater mussels, aquatic insects, and some plant material are also eaten. The

The adult's plastron is pale yellow with black seams.

invasive zebra mussels introduced into northern lakes and some southeastern rivers may provide food but may also decrease the diversity of native food sources.

REPRODUCTION Courtship occurs in both spring and autumn. The male bobs his head in front of the female to determine if she is receptive. Unlike some other map turtle species, males do not have long claws with which to stroke the female's face. Nests are constructed in open, sunny, sandy areas, often not very far from the water. Females lay from one to three clutches of 6–17 eggs each. The eggs hatch in late summer, but hatchlings often remain in the nest over the winter, delaying emergence until the following spring.

The carapace markings are an intricate series of faint yellow lines that resemble a topographic map.

PREDATORS AND DEFENSE Nest predators, including raccoons and fish crows, are the most serious threat to common map turtle populations. Raccoons and larger predators can also kill nesting females.

CONSERVATION ISSUES The common map turtle has likely declined in areas where river water quality has been compromised and freshwater mollusk populations have declined as a result. Given the widespread range of the common map turtle, however, the species seems secure at present.

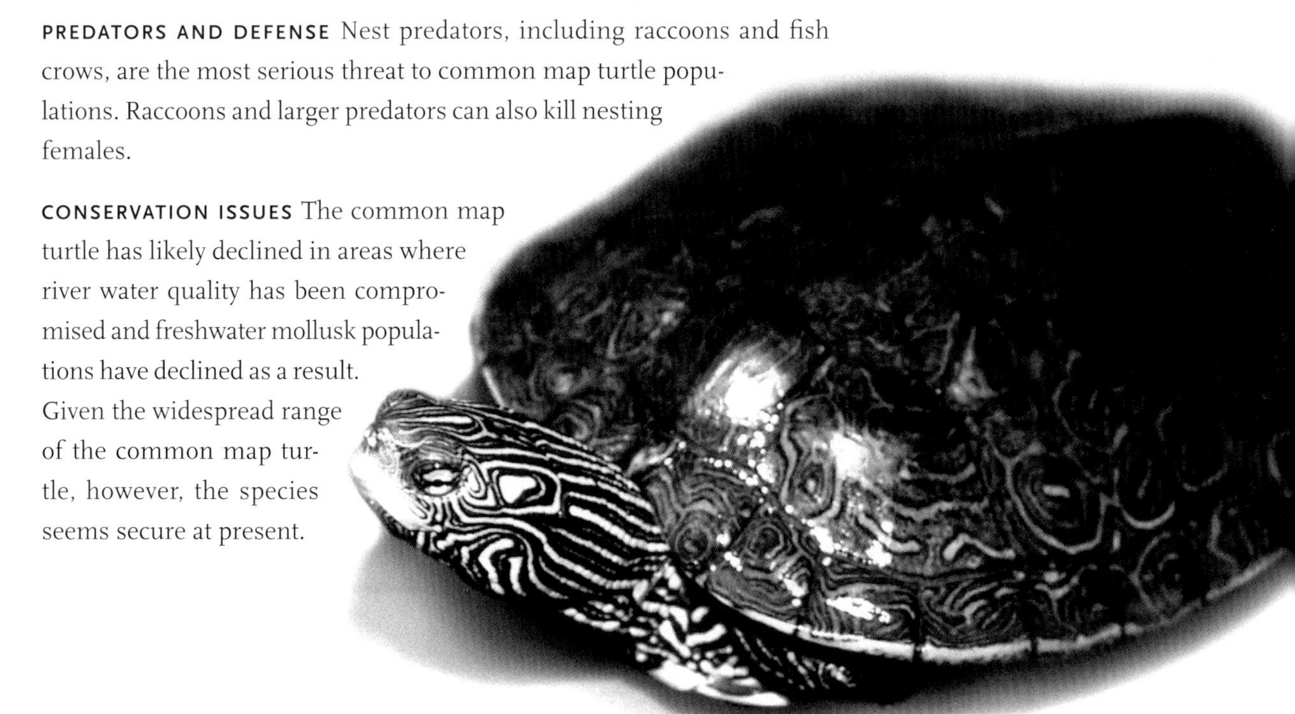

A female Mississippi false map turtle with a yellow crescent behind the eye and white irises.

How do you identify a false map turtle?

DISTINGUISHING CHARACTERS
Yellow vertical bar or crescent behind each eye

AVERAGE SIZE

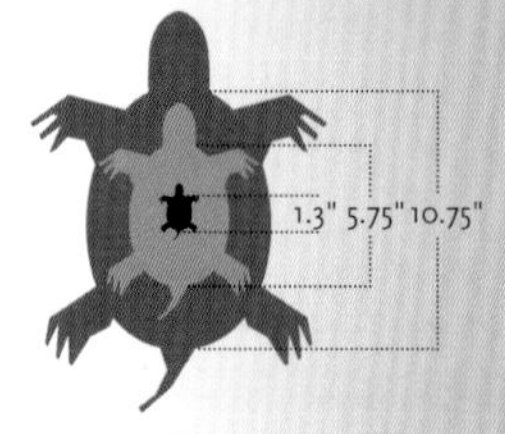

● FEMALE
● MALE
● HATCHLING

CARAPACE SHAPE

FRONT VIEW

SIDE VIEW

TOP VIEW

False Map Turtle *Graptemys pseudogeographica*

DESCRIPTION False map turtles are moderately large turtles, and females are almost twice as large as males. The carapace is olive to greenish brown. The head bears a series of yellow stripes and a heavier mark behind the eye that is a short vertical bar on the midwestern false map turtle (*G. p. pseudogeographica*) and a full crescent on the Mississippi map turtle (*G. p. kohnii*) and prevents the lateral head stripes from intersecting with the eye. The plastron is yellow with brown smudging or a pattern of parallel lines. The iris of the Mississippi map turtle is white, and the eye of the midwestern false map turtle is yellow-brown with a horizontal black bar. Adult males have elongated foreclaws.

VARIATION AND TAXONOMIC ISSUES The group collectively known as the false map turtles has been difficult for turtle biologists to categorize. The Ouachita and Sabine map turtles, which we treat here as full species, have at times been regarded as subspecies of false map turtles. The two southeastern false map turtles—the Mississippi map turtle and the midwestern false map turtle—are treated here as subspecies of a single species, although each has been recognized in the earlier literature as a separate species. The shifting taxonomy is a reflection of the geographic variation that is rampant in this group. The two subspecies can be distinguished by the presence of black spots or half moons on the posterior of some or all

False Map Turtle

Graptemys pseudogeographica

Hatchlings resemble the adults. This individual (above) from the center of the species' range shows intergrade characteristics of both subspecies.

A midwestern false map turtle (left) with a short vertical yellow bar on the head and a horizontal bar through the eye.

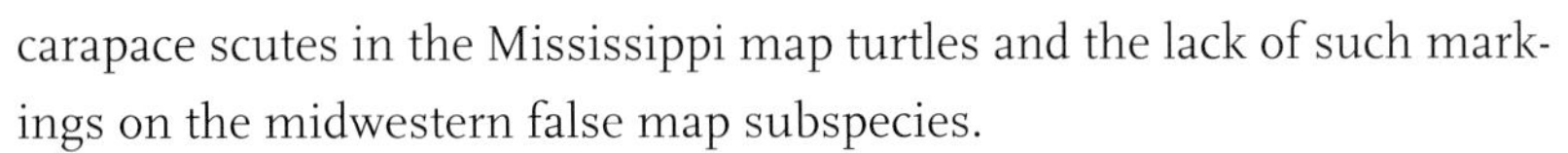

carapace scutes in the Mississippi map turtles and the lack of such markings on the midwestern false map subspecies.

WHAT DO THE HATCHLINGS LOOK LIKE? Hatchlings resemble adults, but the carapace is distinctly keeled. The plastron of hatchling midwestern false map turtles is darker than that of adults, and the pattern on the plastron of Mississippi map turtles is more pronounced in hatchlings than in adults.

CONFUSING SPECIES Ouachita and Sabine map turtles have a distinct yellow spot or blotch behind the eye. Common map turtles have intricate lines on the head, often a very small triangular spot behind the eye, and maplike lines on the carapace.

The head pattern of this false map turtle is intermediate between the two subspecies.

DISTRIBUTION AND HABITAT The false map turtles have a wide distribution that generally includes the Mississippi, Missouri, and Arkansas river drainages. Midwestern false map turtles are found in the middle to upper portions of the Mississippi River drainage, specifically in western Tennessee and western Kentucky. Turtles from regions within Arkansas, Louisiana, and Mississippi best fit the descriptions of the Mississippi map turtle. Intergradation occurs in a wide zone in which individuals show intermediate characteristics. Although typically a species of riverine habitats, false map turtles also frequent river-connected floodplain swamps, oxbows, and sloughs. They occasionally use isolated and seasonal floodplain wetlands.

BEHAVIOR AND ACTIVITY Like all map turtles, false map turtles spend much of their time basking. They have been observed basking on muskrat lodges and may even hibernate within them. Juveniles hibernate in shallow backwaters where they may be more susceptible to winterkill if the water levels drop and expose them to freezing temperatures.

FOOD AND FEEDING The false map turtle is more omnivorous than other map turtles. In the northern part of the range, midwestern false map turtles eat plants, aquatic insects, and mollusks. In the more southern parts of the range, Mississippi map turtles are also omnivorous, but females tend to specialize on mollusks.

REPRODUCTION The male uses his long claws to stroke the sides of the female's face during courtship. Nesting occurs on cloudy days. Females may gather in large numbers in the water near a favorite sandbar or sandbank and nest synchronously on an appropriate day. Each female lays from one to three clutches per year containing 2–8 eggs each. Hatchlings probably leave the nest after hatching and rarely overwinter in it.

PREDATORS AND DEFENSE Nests have been destroyed by raccoons, foxes, skunks, and coyotes, which will also eat hatchlings and may attempt to eat nesting females. Gulls, crows, and even blackbirds and grackles have been reported to eat hatchling false map turtles as they emerge from the nest.

CONSERVATION ISSUES Pollution and channel and floodplain alterations of the Missouri and Mississippi rivers have been cited as causes of this species' decline. Continued monitoring of false map turtle populations will be necessary to track their trends and status.

The Ouachita map turtle's carapace is brownish green with a median keel and blunt knobs or spines on the second and third vertebral scutes.

Ouachita Map Turtle *Graptemys ouachitensis*

DESCRIPTION The Ouachita map turtle is a medium-sized turtle, and females are significantly larger than males. The carapace is brownish green with a median keel and blunt knobs or spines on the second and third vertebral scutes. The raised keel on the spine may form a black stripe that runs the length of the shell. The plastron is yellow with an elaborate black figure. Ouachita map turtles have moderate-sized heads. Up to seven lines on the head reach the eye orbit, and a black horizontal stripe is often visible through the iris of the eye. The lower jaw has three conspicuous spots on the underneath side, and a yellow oval spot or large blotch is present behind each eye. The blotches are separated by lines on the top of the head. Males have elongated foreclaws.

The plastron of a hatchling.

VARIATION AND TAXONOMIC ISSUES Some earlier accounts placed the Ouachita map turtle as a subspecies of the false map turtle.

WHAT DO THE HATCHLINGS LOOK LIKE? Hatchlings resemble the adults, except that they are nearly round. They have a prominent carapace keel and extensive patterning on the plastron.

How do you identify an Ouachita map turtle?

DISTINGUISHING CHARACTERS
Yellow plastron with black figures; three light spots on chin

AVERAGE SIZE

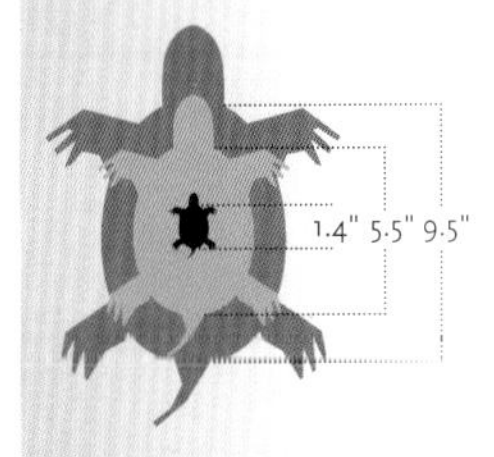

● FEMALE
● MALE
● HATCHLING

CARAPACE SHAPE

FRONT VIEW

SIDE VIEW

TOP VIEW

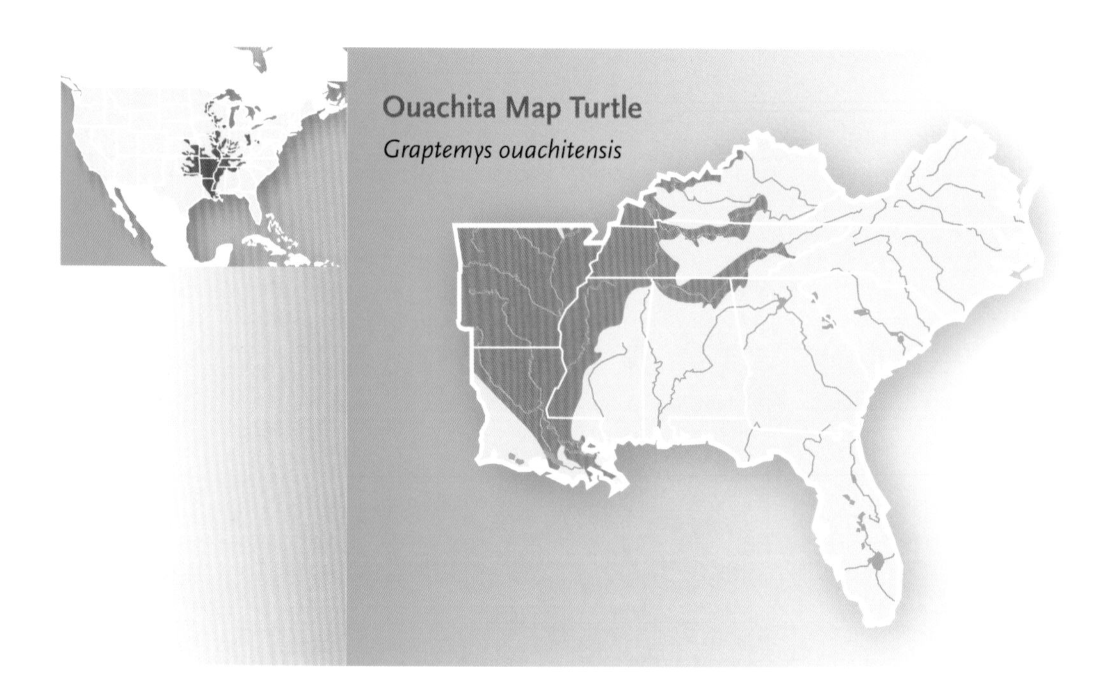

CONFUSING SPECIES Ouachita map turtles can be easily confused with midwestern false map turtles (*G. p. pseudogeographica*), but the latter species lacks the three spots under the chin. The Mississippi map turtle (*G. p. kohnii*), which occupies much of the Ouachita map turtle's range in the Southeast, has a curved crescent behind the eye, a white iris without a stripe through it, and black spots or half moons on the posterior of some or all carapace scutes. Sabine map turtles have a small round spot behind the eye and occur only in the Sabine River system of western Louisiana. The common map turtle lacks prominent head spots or blotches.

DISTRIBUTION AND HABITAT The Ouachita map turtle is found predominantly in the southern Mississippi River drainage, although it extends to northwestern Kentucky in the Ohio River. It occurs in rivers throughout the Ozarks, Ouachita Mountains, Mississippi River Alluvial Plain, and Gulf Coastal Plain. It can be found in both slow- and fast-moving sections of rivers, but prefers areas with submerged vegetation and is occasionally seen in backwater sloughs and floodplain swamps. In Tennessee and northern Alabama, the Ouachita map turtle occurs in the Tennessee River drainage, where it lives in the slower sections of rivers as well as in reservoir impoundments.

This individual has unusually bold yellow head markings.

BEHAVIOR AND ACTIVITY Ouachita map turtles bask frequently on log snags, often in large numbers. They

A female digs a nest.

A brightly patterned male.

The plastral pattern fades with age.

are wary and difficult to approach. Little else is known of the ecology of this species.

FOOD AND FEEDING Ouachita map turtles feed on both plants and insects; juveniles may be more carnivorous than adults. Presumably, the larger females eat snails and mussels, but perhaps in lesser amounts than other map turtles.

REPRODUCTION The male vibrates his foreclaws in front of the female to induce her to mate. Nesting is not well documented in this species, although females are thought to lay one to three clutches of 3–13 eggs in the summer. In Arkansas, these turtles nest on sandbars.

PREDATORS AND DEFENSE Mammalian nest predators, such as raccoons, are undoubtedly a threat. Hatchlings must elude herons, fish crows, gulls, and large fish.

CONSERVATION ISSUES The wide range of the Ouachita map turtle and its occurrence in many different streams in several geographic regions bodes well for its future. Its minimal dependence on mollusks, which have suffered in most southeastern streams, may also work in its favor. Nevertheless, populations should be monitored.

The carapace is brown to olive green. This individual is an adult female.

How do you identify a Sabine map turtle?

DISTINGUISHING CHARACTERS
Small spot behind each eye; many thin, faint stripes on neck; males with very narrow head; found only in the Sabine River drainage

AVERAGE SIZE

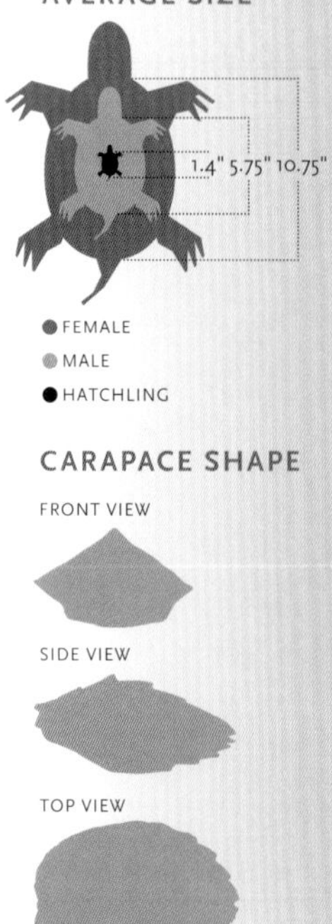

Sabine Map Turtle *Graptemys sabinensis*

DESCRIPTION These small to medium-sized turtles are among the smallest of the map turtles. The carapace is brown to olive green, often with thin yellow concentric lines on each carapace scute. The keel on the carapace is prominent, especially on males, and the posterior of the shell is serrated. The plastron is basically plain white or cream colored, with darker markings in the seams. Numerous thin yellow neck stripes reach the back of the eye; a small spot is also present behind the eye. The females weigh three to four times more than the males but do not have a broad head. Males have a very narrow head and slightly elongated foreclaws.

VARIATION AND TAXONOMIC ISSUES Sabine map turtles are similar to Ouachita map turtles, but some turtle biologists recognize them as a distinct species. Sabine map turtles have a small roundish or oval spot behind each eye, and Ouachita map turtles have a large blotch.

Males have a very narrow head.

WHAT DO THE HATCHLINGS LOOK LIKE? Sabine map turtle hatchlings are similar to the adults but have a nearly round shell. The carapace keel is more prominent and the carapace edges are serrated.

Sabine Map Turtle
Graptemys sabinensis

The plastron (left) and the carapace (right) of a hatchling.

CONFUSING SPECIES The Sabine map turtle's range overlaps with that of the Mississippi map turtle (*G. p. kohnii*), but the latter has a distinct yellow crescent behind the eye on each side of the head.

DISTRIBUTION AND HABITAT Sabine map turtles occur in the Sabine, Calcasieu, and Mermentau river drainages of western Louisiana in the Gulf Coastal Plain.

BEHAVIOR AND ACTIVITY Sabine map turtles bask on logs and snags. As is true of most map turtles, they are wary and drop into the water at the slightest hint of danger.

FOOD AND FEEDING The small, narrow-headed males eat aquatic insects such as caddisfly larvae and some algae that they pick from submerged logs. Females are said to eat filamentous algae in large amounts and occasionally snails, dead fish, crayfish, and aquatic insects.

Did you know?

Graptemys *is the most species-rich genus of turtles in the Southeast.*

The sabine map turtle has a small round spot behind the eye.

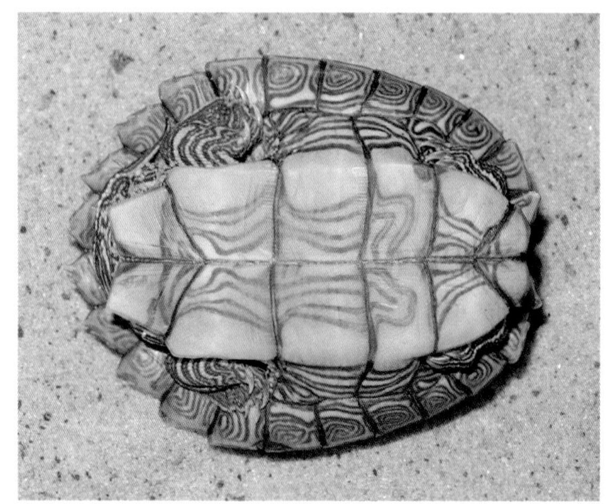

The plastron is plain white or cream colored with concentric black lines.

REPRODUCTION Very little is known about the reproductive habits of the Sabine map turtle. Limited data indicate that females nest in late spring and summer on sunny sites, probably on sandy riverbanks. The average clutch size is about 6.

PREDATORS AND DEFENSE Nests are likely to be disturbed by raccoons, fish crows, and other predators. Hatchlings must hide from herons, gulls, and fish such as largemouth bass and gar. Alligators are also possible predators.

CONSERVATION ISSUES The Sabine map turtle has a limited distribution, and thus is vulnerable to pollution of the rivers in which it occurs. Historical impacts from logging, agriculture, oil exploration, and reservoir construction have all affected the current distribution of the turtles or had serious environmental impacts on Sabine River habitats.

Conspicuous black knobs are present on this black-knobbed sawback.

How do you identify a black-knobbed map turtle?

DISTINGUISHING CHARACTERS
Olive shell with prominent black knobs along midline of carapace

AVERAGE SIZE

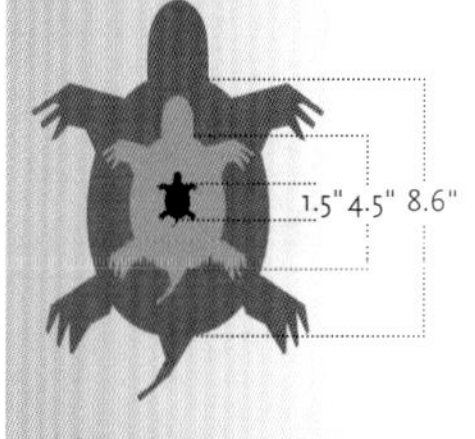

● FEMALE
● MALE
● HATCHLING

CARAPACE SHAPE

FRONT VIEW

SIDE VIEW

TOP VIEW

Black-knobbed Map Turtle *Graptemys nigrinoda*

DESCRIPTION The black-knobbed map turtle is a medium-sized turtle. Females are larger than males, but the sexual dimorphism is not as great as in some species. The black-knobbed map turtle is distinctive in having the most pronounced carapace keel of any of the map turtles. The vertebral projections appear as black knobs above the spine, especially on the second and third vertebral scutes, and are taller on males than on females. The carapace is green to brown, with hollow semicircular or circular yellow rings on the carapace scutes. The plastron is variably colored, with some individuals showing little black patterning and others having the majority of the plastron pigmented. The head is narrow with yellow striping; two to four head stripes intersect with each eye. Males have elongated foreclaws. The black-knobbed map turtle, the yellow-blotched map turtle, and the ringed map turtle are often referred to as the "sawbacks" because of their prominent dorsal knobs.

Typical head pattern found on the northern subspecies.

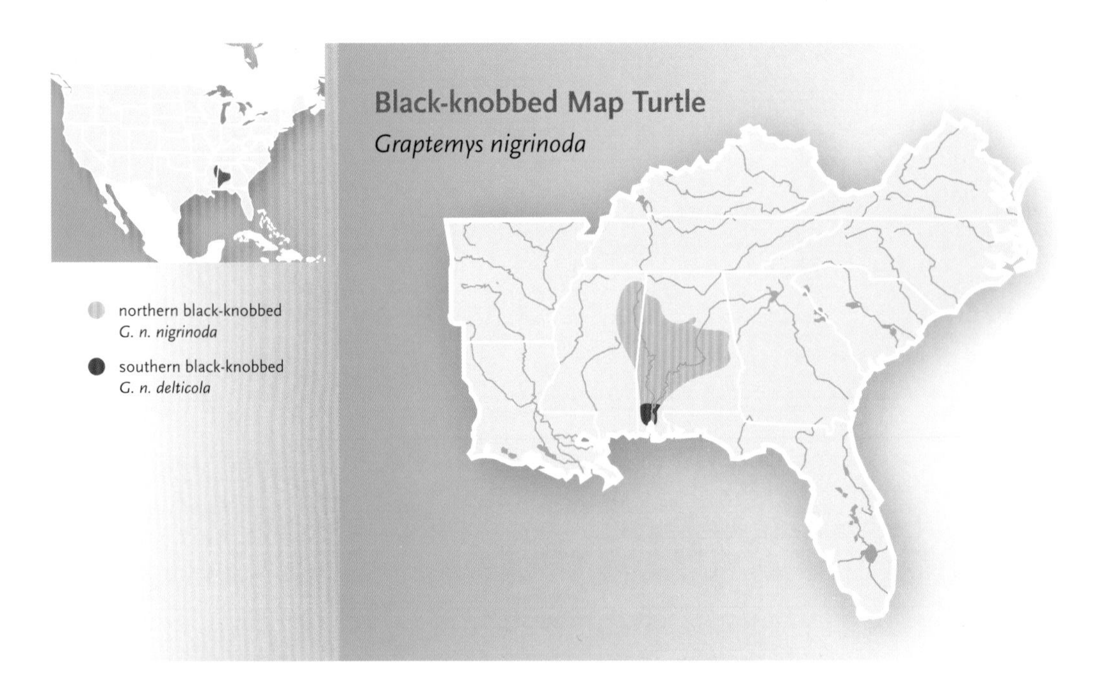

Hatchlings resemble adults but have more prominent dorsal knobs (top) and extended marginal scutes (bottom).

VARIATION AND TAXONOMIC ISSUES Two subspecies are recognized. The northern black-knobbed map turtle, *G. n. nigrinoda*, is lighter skinned, and the plastron is yellow with black marking within the seams. The southern black-knobbed map turtle, *G. n. delticola*, has darker skin, and the black pattern extends beyond the seams and covers much of the plastron.

WHAT DO THE HATCHLINGS LOOK LIKE? Hatchlings resemble adults but with dorsal knobs of proportionately greater height than in adults of either sex.

CONFUSING SPECIES Alabama map turtles have large blotches on the sides of the head, and females have an enlarged head. Slider turtles also have yellow blotches on the sides of the head. Alabama red-bellied cooters and river cooters do not have a keel on the carapace. The prominent vertebral knobs should prevent confusion with any other turtle within the species' range.

DISTRIBUTION AND HABITAT The northern black-knobbed map turtle occurs in the Alabama, Tombigbee, Coosa, Black Warrior, Tallapoosa, and Cahaba rivers, from the Fall Line southward onto the Gulf Coastal Plain. Sandy-bottomed, flowing river sections are the preferred habitats. The southern black-knobbed map turtle appears restricted to the lower Mobile Bay drainage, where it inhabits slower-moving waterways that connect the stream courses and lakelike habitats of that region.

BEHAVIOR AND ACTIVITY The basking habit is well developed in black-knobbed map turtles. They prefer logs and branches that are surrounded

A male basks on a log snag in the Alabama River.

The southern black-knobbed map turtle (left) has a darker plastron than the northern subspecies (right).

by water but will bask on riverbanks during periods of flooding. They show strong fidelity to individual basking sites and often spend the nights on underwater sections of the same basking snags. They are wary of anyone approaching. Black-knobbed map turtles do not leave the river except to nest. They are shy and do not attempt to bite.

FOOD AND FEEDING Black-knobbed map turtles are most likely omnivorous. Males consume aquatic insects as well as terrestrial insects that fall into the water. Females may consume plant material as well as snails and freshwater mussels. Southern black-knobbed map turtles are reported to eat barnacles and blue crabs, indicating that the interface between fresh

This female black-knobbed sawback is much larger than the male.

and salt water in the lower Mobile Bay ecosystem is included in their range.

REPRODUCTION Although data on age at maturity are difficult to obtain, it appears that black-knobbed map turtle males mature at approximately age 5 and females at age 9. Courtship may include behavior in which the male bobs his head at the female under water. Nesting occurs during late May to mid-July, and the females may lay multiple clutches of 3–7 eggs. They dig nests after dark in open, sandy areas with scattered clumps of grass. Hatchlings have been reported to emerge from mid-August to early October.

PREDATORS AND DEFENSE Fish crows, common crows, raccoons, and armadillos destroy many nests. Perhaps by nesting at night black-knobbed map turtles limit the ability of crows to find their nests, although they certainly find many of them. Alligators likely pose occasional threats to adult turtles, especially the smaller males.

CONSERVATION ISSUES The black-knobbed map turtle has a fairly extensive range within central and southern Alabama. Camping and intensive day use of river sandbars during the nesting season may affect nesting success. Fire ants, invasive vegetation, and raccoons may collectively reduce hatching success with long-term consequences on population size and stability. Adults are occasionally found with injuries attributable to motorboat propellers. Illegal collecting may also be a problem in some areas. Of all the map turtles in the Southeast, however, the black-knobbed map turtle is currently one of the most secure.

A yellow-blotched map turtle with characteristic yellow blotches on the carapace scutes.

Yellow-blotched Map Turtle

Graptemys flavimaculata

DESCRIPTION Yellow-blotched map turtles are medium-sized turtles. Females are larger than males, but the size difference is not as great as in some species of map turtles. They have a pronounced carapace keel, with raised knobs that resemble dark fins above the spine, especially on the second and third vertebral scutes. These projections are more distinct in males than in females, and even more so in juveniles. The carapace is brown, with a solid yellow blotch—rarely a hollow yellow ring—in the center of each scute. The plastron is pale yellow to white with black patterning along the scute seams. The head is narrow with yellow striping, and with two to four head stripes intersecting with the eye. A small blotch, usually rectangular, extends behind each eye. Males have elongated foreclaws. The yellow-blotched map turtle, black-knobbed map turtle, and ringed map turtle are often referred to as the "sawbacks" because of their prominent dorsal knobs.

Occasionally, yellow-blotched map turtles can have hollow rings instead of solid blotches on the carapace.

How do you identify a yellow-blotched map turtle?

DISTINGUISHING CHARACTERS
Yellow to orange solid blotch on each carapace scute

AVERAGE SIZE

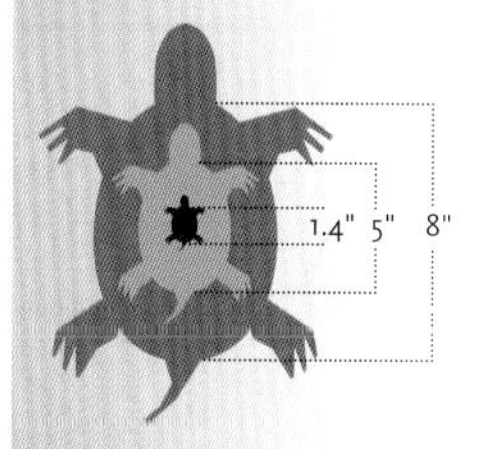

FEMALE
MALE
HATCHLING

CARAPACE SHAPE

FRONT VIEW

SIDE VIEW

TOP VIEW

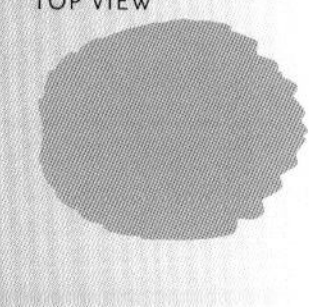

Hatchlings.

VARIATION AND TAXONOMIC ISSUES No subspecies have been described, and no geographic variation is apparent.

Yellow-blotched map turtles are often referred to as "sawbacks." The egg tooth is still present on this newly emerged hatchling.

WHAT DO THE HATCHLINGS LOOK LIKE? Hatchlings resemble adults but are more strongly keeled.

CONFUSING SPECIES The yellow blotches on the carapace separate most individuals of this species from all other turtles. River cooters lack a dorsal keel, and the posterior marginals of the carapace are generally smooth. Gibbons' map turtles also have a strongly keeled carapace, but have large blotches on each side of the head instead of stripes.

DISTRIBUTION AND HABITAT Yellow-blotched map turtles are endemic to the Pascagoula River and its major tributaries, the Leaf and Chickasawhay rivers, within the Gulf Coastal Plain of Mississippi. The appropriate habitat includes numerous log snags, exposed stumps, and brush piles for basking. Sandbars that are elevated above normal summer floods provide critical nesting habitat.

BEHAVIOR AND ACTIVITY Yellow-blotched map turtles are most often observed as they bask on log snags and among branches of overhanging or fallen trees. Juveniles often bask within the network of branches, perhaps as protection from aerial predators or large fish. At night, juveniles and adult males will sit underwater on the submerged portions of the same logs they basked on during the day. Bumping the log at night with a boat will bring disoriented turtles to the surface.

FOOD AND FEEDING The small males feed primarily on aquatic insects that attach themselves to log snags. Adult females feed on snails, small native mussels, introduced Asiatic clams, and perhaps less frequently on tadpoles and aquatic plants.

REPRODUCTION The male courts the female by stroking her face with his long front claws. Females generally nest on large sandbars between early May and early August, although occasionally nests are found on small sandbars and steeply eroded riverbanks. Clutches contain 3–9 eggs (aver-

Yellow-blotched Map Turtle
Graptemys flavimaculata

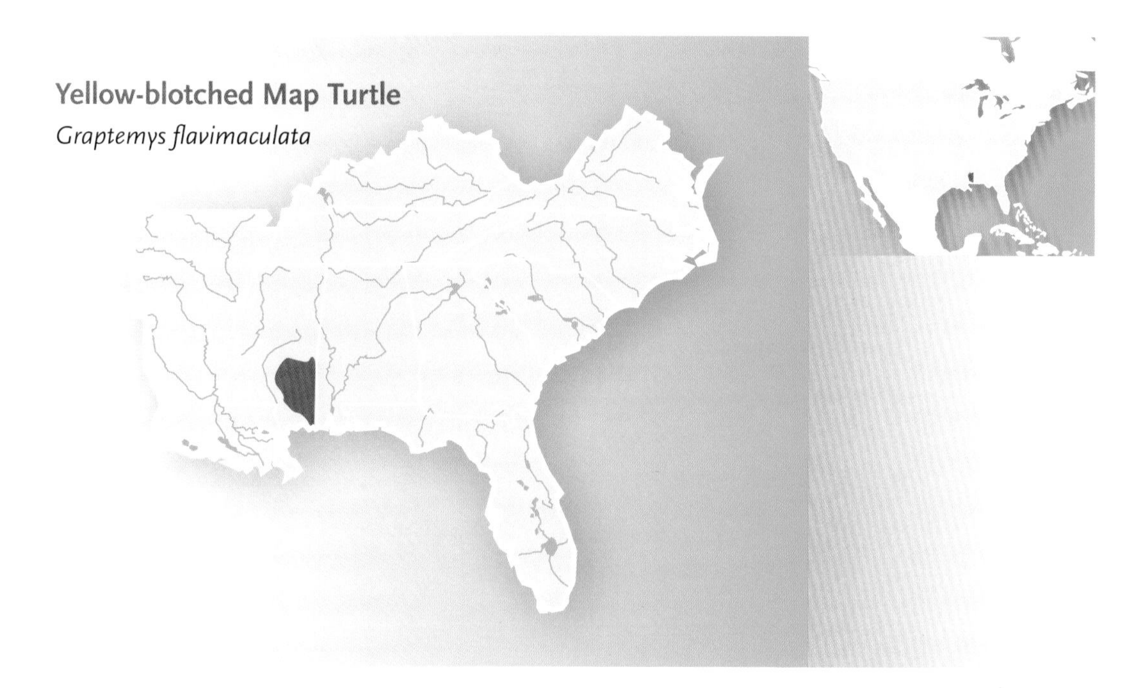

age = 5). Nesting females are easily disturbed by human activities such as camping and picnicking. Declines in reproduction have been linked to frequent human disturbance of the nesting beaches. Hatching probably occurs from August through September.

A distinctive black pattern is found on the plastron. This individual is a male.

PREDATORS AND DEFENSE Fish crows, raccoons, and feral pigs dig up many nests. Introduced fire ants kill hatchlings before they can emerge from the nest. Adult females reach sizes that few native predators besides an alligator are likely to eat, although bite marks have been observed on turtles that obviously survived an attack. Adult turtles may bite when handled but generally withdraw into the shell.

CONSERVATION ISSUES The Pascagoula River has been polluted by wood pulp and turpentine processing, raw sewage, and other processes, and aquatic insect and mollusk populations have declined as a result. Sandbars used for nesting have also been affected by dams and flooding. Introduced plants carpet once-open sandbars, and recreational use of sandbars by humans during the peak nesting period disturbs nesting females and leads to fewer nests placed in optimal habitat. Illegal shooting of basking turtles for sport is also a persistent problem. The species was listed as Threatened under the federal Endangered Species Act in 1991 due to declining populations and habitat conditions. A comprehensive survival plan will involve protection of nests, increased survivorship of adults, restoration of riverine and nesting habitats, and local education and law enforcement.

A female showing typical hollow red rings.

How do you identify a ringed map turtle?

DISTINGUISHING CHARACTERS

Yellow to red rings on each carapace scute

AVERAGE SIZE

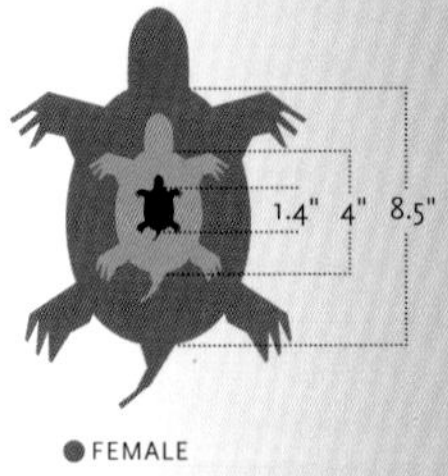

● FEMALE
● MALE
● HATCHLING

CARAPACE SHAPE

FRONT VIEW

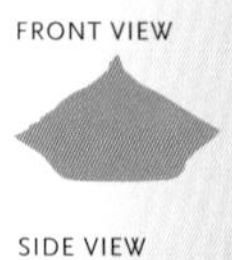

SIDE VIEW

TOP VIEW

Ringed Map Turtle *Graptemys oculifera*

DESCRIPTION Ringed map turtles are medium-sized turtles. Females are larger than males, but the sexual dimorphism is not as great as in some species. Ringed map turtles have a pronounced midline keel on the carapace, and the vertebral projections appear as dark fins above the spine, especially on the second and third scutes. The projections are more distinct in males than in females. The carapace is brown, and a distinctive hollow red or yellow circle or ocellus occupies the center of each costal scute. The plastron is pale yellow to white with black patterning along the seams of the scutes. The head is narrow, with yellow striping. Two to four head stripes intersect with the eye. A small blotch, usually rectangular, extends behind each eye. Males have elongated foreclaws. The ringed map turtle, yellow-blotched map turtle, and black-knobbed map turtle are often referred to as the "sawbacks" because of their prominent dorsal knobs.

VARIATION AND TAXONOMIC ISSUES No subspecies are recognized, and no variation has been noted.

Yellow-red rings on carapace scutes of a ringed map turtle.

A hatchling.

WHAT DO THE HATCHLINGS LOOK LIKE? The hatchlings resemble the adults; the dorsal spines are tall and tipped with black.

CONFUSING SPECIES River cooters, which are often seen basking with ringed map turtles, have distinct yellow stripes on the sides of the head and lack a dorsal keel and red rings on the scutes. Gibbons' map turtles have large yellow blotches on each side of the head. The red-eared slider turtles of the region have broad red stripes on the head behind each eye and conspicuous light bars perpendicular to the spine on the costal scutes of the carapace.

DISTRIBUTION AND HABITAT Ringed map turtles are found only in the Pearl River and its tributaries, most notably the Bogue Chitto River in Mississippi and Louisiana. The Pearl River is relatively wide and swift flowing in the lower reaches, with large sandbars and a predominantly sandy bottom. Log snags and piles of trees and branches form basking and hiding locations. Ringed map turtles also occur where the river approaches the Gulf of Mexico and the current slows and becomes swamplike, but they do not enter brackish water. They are never seen on land except when the females are nesting on sandbars, and they do not inhabit isolated seasonal wetlands or ponds.

The steep, keeled shell is noticeable in this juvenile.

Sandbars above the flood line with sparse vegetation are favored nesting sites on the Pearl River.

BEHAVIOR AND ACTIVITY Ringed map turtles are wary during basking. They prefer to sit on logs and intertwined branches that are isolated from the shore and drop off at the slightest disturbance, disappearing into the often murky and churning waters.

Typical black patterning on the plastron of a male ringed map turtle.

FOOD AND FEEDING Both sexes appear to feed primarily on aquatic insects such as dragonfly and damselfly nymphs, caddisflies, aquatic beetles, and blackfly larvae. Small snails and some algae are also consumed. Freshwater mussels may be eaten less commonly.

REPRODUCTION Courtship has not yet been observed. Females nest during the day, depositing eggs on sandbars from May to mid-July; most eggs hatch in August. Clutches contain 1–10 eggs (average = 4), and most females lay one clutch per year. Hatchlings are not believed to overwinter in the nest.

PREDATORS AND DEFENSE Nest predators include armadillos, raccoons, fish crows, and imported fire ants. Ringed map turtles generally rely on their wariness to escape predators. Hatchlings are vulnerable to wading birds such as herons and egrets. Otters may occasionally eat juveniles or small males. Natural predators of adults certainly include the occasional alligator.

CONSERVATION ISSUES Ringed map turtles are declining for many reasons. The species is naturally vulnerable because it is restricted to a single major river system. Sewage, industrial contaminants, and agricultural runoff are poisoning the species' invertebrate food base. Channelization has altered the flow in many river stretches, affecting river bottoms, basking site availability, and the natural processes that create sandbars. Human use of sandbars for camping and picnicking may affect nesting success. Restoring natural flow conditions and not treating our rivers as sewers will alleviate some of the environmental problems faced by the ringed map turtles and others living in similar habitats. In 1986 the species was listed as Threatened on the federal endangered species list as a result of declining populations and habitat conditions.

BRACKISH WATER TURTLES

A terrapin on the bank of a tidal creek at low tide.

Diamondback Terrapin *Malaclemys terrapin*

DESCRIPTION Diamondback terrapins are among the most variable of North American turtles. Adults typically have a light to dark gray shell with lighter-colored concentric rings, and a gray head and limbs with darker spots or blotches, but an array of colors and patterns can be found throughout the range. Even a single population can include some individuals with dark gray to almost black shells and body parts, others with gray bodies and shells with orangish rings on a light gray or greenish background, and numerous variations in between. The plastron has no hinge and ranges from light yellowish to almost black, often in accord with the general body color of the individual. The feet are webbed, providing the strong swimming abilities required in an environment of tidal fluctuations and powerful currents. Adult females are much larger than males and develop a disproportionately larger head and jaws.

VARIATION AND TAXONOMIC ISSUES Six of the seven subspecies of terrapins occur along coastal areas of the southeastern states from Virginia to Louisiana. The northern (*M. t. terrapin*) and Carolina (*M. t. centrata*) subspecies are generally described as having no knobs down the center of the carapace; the other subspecies do have knobs. Some females of the Carolina subspecies seem to have a larger head than those of the northern

How do you identify a diamondback terrapin?

DISTINGUISHING CHARACTERS
Only turtle residing in brackish water; pattern very variable but skin usually gray

AVERAGE SIZE

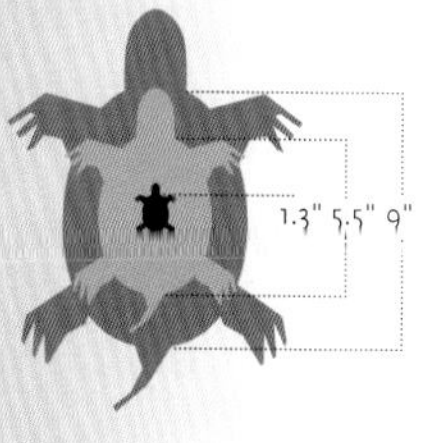

● FEMALE
● MALE
● HATCHLING

CARAPACE SHAPE

FRONT VIEW

SIDE VIEW

TOP VIEW

Diamondback terrapins are highly variable in color and pattern. Each of these pictures was taken in the identified range of that subspecies.

northern diamondback, *M. t. terrapin* (left); Carolina diamondback, *M. t. centrata* (middle); Florida east coast diamondback, *M. t. tequesta* (right), with barnacles

Typical habitats include tidal creeks in salt marshes, estuaries of rivers, and the margins of bays where freshwater meets saltwater.

Diamondback Terrapin
Malaclemys terrapin

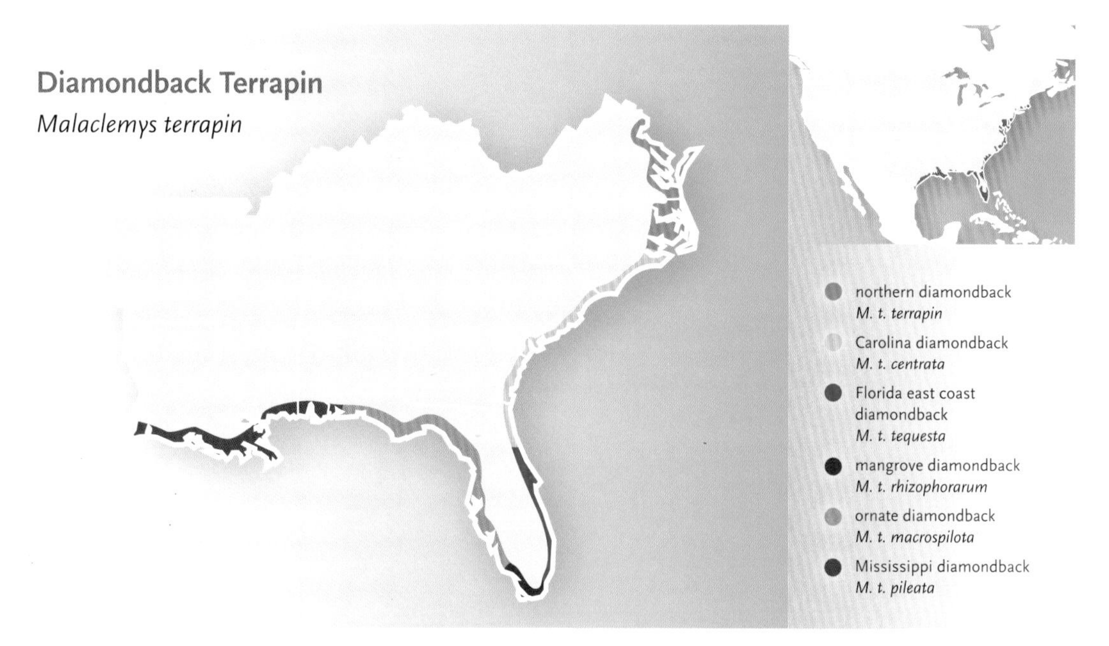

mangrove diamondback, *M. t. rhizophorarum* (left);
ornate diamondback, *M. t. macrospilota* (middle);
Mississippi diamondback, *M. t. pileata* (right)

subspecies. The Florida east coast subspecies (*M. t. tequesta*) often lacks concentric circles on the carapace scutes, and the mangrove subspecies (*M. t. rhizophorarum*) has bumps on the dorsal keel and streaks on the neck rather than spots. The ornate subspecies (*M. t. macrospilota*) typically has large yellow or orange blotches on the carapace in the center of each scute. The Mississippi subspecies (*M. t. pileata*) is characteristically dark with yellow or orange tinting on the marginal scutes. As was noted above, however, the color patterns and appearance can vary widely within a single population or over only a few miles of salt marsh coastline. Terrapins have long been farmed for food, and the capture, subsequent release, and mixing of terrapins from many regions may have contributed to the extensive variation seen today.

Extreme variability in pattern and color can be seen in diamondback terrapins throughout much of their geographic range. All of the variations shown here can be seen in a single population at Kiawah Island, South Carolina.

WHAT DO THE HATCHLINGS LOOK LIKE? Baby terrapins are characteristically gray or brownish above with a lighter-colored plastron. The carapace may be knobbed down the center, and the scutes may be a solid color or have spots or circles that contrast with the base color. Babies may appear slightly more round than the adults.

CONFUSING SPECIES Identification is not usually an issue with this species, because other turtles seldom enter its brackish water habitats, and terrapins rarely venture into strictly freshwater habitats. The lack of stripes or other yellow markings on the head, the absence of hinges on the plastron, and the relatively smooth marginal scutes without serrations at the rear of the carapace separate this species from others whose habitats might be contiguous. Terrapins resemble map turtles, which are their close relatives.

Hatchling terrapins are characteristically gray or brownish above with a lighter-colored plastron.

DISTRIBUTION AND HABITAT Diamondback terrapins have historically been found in coastal areas of all the southeastern states. They live in neither marine nor freshwater habitats, but thrive instead in brackish waters generally avoided by other North American turtles. Typical habitats include tidal creeks in salt marshes, estuaries of rivers, and the margins of bays where fresh water meets salt water. In South Florida, diamondback terrapins live in tropical mangrove swamps. Terrapins can survive in fresh water, but no natural populations exist in purely freshwater habitats.

BEHAVIOR AND ACTIVITY Like other southeastern turtles, diamondback terrapins are active throughout the warm portion of the year. They are most often seen in the spring and fall, hibernating during cold winter weather and becoming less active during hot weather in the summer. The period of winter inactivity may be very short or almost nonexistent in southern Florida populations. Terrapins are most likely to be seen swimming with the head poking above the water's surface. They can commonly be seen in tidal creeks or coastal rivers at low tide but are less obvious during high tides, when large predators such as sharks and other large fish are more likely to enter their habitat. Turtle biologists suspect that terrapins spend high tide hidden in the grasses and mud of the salt marsh flats, where they feed on periwinkle snails. Many a crabber has pulled up a crab trap and discovered a terrapin inside. During the nesting season, adult females are frequently encountered crossing causeways and other roads that separate the salt marsh from suitable nesting habitat. In the mangrove swamps of South Florida, terrapins bask on mangrove limbs and the sides of tidal creeks, especially during cool, sunny days in spring and fall, and bury themselves in the mud among the mangrove roots at low tide.

During the nesting season, adult females leave the salt marsh to nest in sandy dunes, often having to cross roads along the way.

Female terrapins have wider heads than males and can crush periwinkle snails, a favorite food.

Most terrapins are docile and do not bite if they are held, but a few will do so if given the chance, and given their massive head and jaw muscles, it pays to be careful when handling them.

FOOD AND DIET Terrapins are almost strictly carnivorous, eating a variety of shellfish (such as periwinkle snails) and crustaceans (such as soft-shell or hard-shell blue crabs). The larger females typically eat animals that are appreciably larger than those eaten by the small-headed males. They also eat dead fish but are probably not very good at catching live fish. Hatchlings have been seen eating tiny fiddler crabs.

REPRODUCTION The life cycle of the diamondback terrapin is typical of most North American species of freshwater turtles. Mating occurs in the water in early spring, and possibly in autumn before cold weather sets in, and aggregations of dozens to more than a hundred individuals have been seen in a single area. Females lay one or two clutches of 4–18 eggs (average = 7) in spring and summer. The nests are constructed on land on sand dunes, on small hammocks in salt marshes, or in terrestrial areas

The female (above) has a larger head and body than the male (below).

peripheral to brackish habitats and above the high-tide mark. Most baby terrapins hatch and leave the nest in late summer.

Where juvenile terrapins live during the years after hatching and before they enter the tidal creeks and rivers at around age 4 is unknown. A few 3-year-old terrapins have been captured by biologists, but turtles 2 years old and younger are almost never seen. They may hide in tidal rack and among grasses on shallow mud flats, where wading birds, large fish, and crabs are less likely to find them.

Did you know?

Only one North American turtle is found exclusively in brackish water: the diamondback terrapin.

PREDATORS AND DEFENSE Raccoons, foxes, crows, and gulls are known nest predators. Near beach areas, ghost crabs will eat eggs or hatchlings. Potential predators on juveniles and small adults include wading birds and large predatory fish such as sharks. Even adults can fall prey to larger sharks, dolphins, and bald eagles. When on land, nesting females are susceptible to predation by raccoons.

CONSERVATION ISSUES Diamondback terrapins have declined considerably throughout their geographic range and continue to do so in many, if not most, regions. Their value as a dietary delicacy has dramatically reduced populations in some areas. Although consumption in North America has declined, demand for terrapins in Asian markets has increased in recent years. Capture and drowning in recreational and possibly commercial crab traps and high road mortality of females and hatchlings have contributed to their decline. The most effective approach to conservation of diamondback terrapins has been at the local level through a combination of regulations (e.g., requiring turtle excluder devices on crab traps), public education and awareness (e.g., posting terrapin crossing signs on roads), and fencing to prevent nesting females from gaining access to roads.

MARINE TURTLES

Marine turtles inhabit the offshore coastal waters and tidal estuaries of the southeastern states, and some nest on their ocean beaches. These "sea" turtles spend their entire lives immersed in the marine environment. The few people lucky enough to see them usually encounter nesting females on nighttime beach walks in the summer or find a dead turtle washed ashore on a beach.

Seven species of marine turtles are known worldwide, and we are fortunate to have five of them as residents or occasional visitors. Loggerhead and Kemp's ridley sea turtles frequent our tidal estuaries and the Intracoastal Waterway during the summer months. Boaters may glimpse a large head surfacing for a breath before the turtle submerges again to search for crabs and shellfish on the bottom. Unfortunately, loggerheads that become trapped in shrimp-trawling equipment and drown are often found stranded (i.e., dead) on Atlantic and Gulf coast beaches. Many concerned shrimpers have installed turtle excluder devices (TEDS) to minimize mortality to marine turtles, yet deaths still occur. Snorkelers on the coral reefs off the Florida Keys often have the opportunity to view green and hawksbill sea turtles. These species are primarily tropical, preferring the warmer waters of the Caribbean, although greens occasionally venture up the Atlantic Coast as far north as Massachusetts, and hawksbills have been seen off the coast of Virginia during the summer months. Green sea turtles feed on aquatic plants, and hawksbills specialize on sponges.

Did you know?

The size and pattern of a female sea turtle's tracks in the sand that lead to a nest provide evidence of what species laid the eggs.

The giant leatherback sea turtles range throughout the world's oceans and even visit Arctic waters, where their large mass may help insulate them and allow them to spend extended periods in the cold waters. They feed exclusively on jellyfish and yet still grow rapidly and become at least twice as large as any of the other marine turtles. Unfortunately, leatherbacks, too, are often found stranded on ocean beaches of the Southeast after swallowing discarded plastic bags that clog their digestive tract, presumably mistaking the bags for jellyfish.

This Kemp's ridley is returning from a nesting site.

Green turtles and leatherbacks nest on Florida beaches, the former species in substantial numbers, and a few individuals nest in Georgia and South Carolina. The nearest nesting sites for hawksbills are generally the Caribbean islands. Kemp's ridley sea turtles nest on two beaches in the Gulf

Adult females are at risk from shrimp nets, which affect their ability to return to nesting beaches in successive years.

Baby loggerhead sea turtles will head straight for the ocean after leaving the nest.

of Mexico and nowhere else. Loggerhead sea turtles nest commonly on southeastern ocean beaches from Virginia to Florida and the Gulf Coast. Loggerheads nest at night, and lucky early-morning beach-goers may find their tracks during the spring and summer months. Hatchlings also emerge under the cover of darkness during late summer and race to the surf, providing another opportunity for people in the Southeast to experience and appreciate marine turtles. Nesting females and hatchlings are the only live sea turtles seen out of water naturally in the Southeast, although greens and hawksbills sometimes bask along shorelines in other parts of the world. All sea turtles are protected under the U.S. Endangered Species Act.

A sea turtle nesting on a beach at night north of Florida will most likely be a loggerhead.

Did you know?

Females of many species of turtles can produce more than one clutch of eggs per year. Some freshwater turtles can nest up to three times, and sea turtles can nest up to seven times in a given year.

Loggerhead *Caretta caretta*

The loggerhead is the most commonly seen marine turtle along the Atlantic and Gulf coasts of the Southeast, and is the only sea turtle that commonly nests on Atlantic beaches north of Florida. It is a rather large turtle, measuring up to 47 inches in carapace length, and can weigh up to 440 pounds. The carapace is longer than wide and has three bridge scutes connecting the plastron to the carapace. It is brown to reddish brown in color.

Kemp's Ridley *Lepidochelys kempii*

Kemp's ridley is the smallest of the Southeast's marine turtles and the one with the smallest worldwide distribution. The carapace is nearly round, up to 27 inches long, and ranges in color from gray to dark gray-green. Four bridge scutes are present between the carapace and plastron. Adults weigh about 100 pounds. The species is most commonly found in the Gulf of Mexico and historically nested on only one beach in Mexico. Recent conservation efforts have established a new nesting beach in south Texas. During the summer months, juveniles are frequently observed in the Chesapeake Bay in Virginia, an important feeding area.

A Kemp's ridley swims in shallow water.

A green sea turtle.

Green *Chelonia mydas*

Green sea turtles are most common along the southern Florida coasts, but they range offshore of all the coastal southeastern states during the summer months. As their common name indicates, these turtles are usually dark gray-green. Green turtles are the second largest marine turtle, with carapace length up to 49 inches and weight up to 506 pounds. One pair of scales is present between the eyes on top of the head. The jaws are serrated, a useful adaptation for shredding aquatic plants.

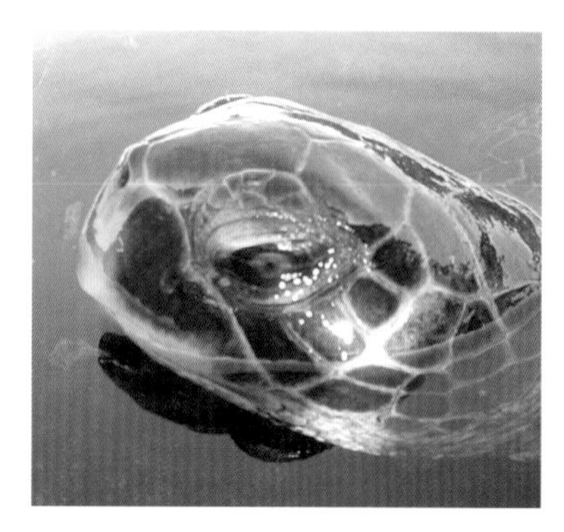

The green sea turtle (left) has only a single pair of scales between the eyes. The hawksbill (right) has two pairs.

Hawksbill *Eretmochelys imbricata*

The hawksbill is a tropical species rarely seen along southeastern coastlines. Adults can measure up to 38 inches in carapace length and weigh up to 187 pounds. The carapace is longer than wide and tapers to a point posteriorly. The color is often a mottling of browns, tans, and yellows. The shell was used historically for "tortoise shell" jewelry, ornaments, combs, and other items. The hawksbill turtle has a distinctly pointed face and an overbite of the upper jaw. Two pairs of scutes are present between the eyes on top of the head.

The hawksbill is rarely seen along southeastern coastlines.

Leatherback *Dermochelys coriacea*

The leatherback is the largest turtle in the world and the reptile with the widest geographic range. Adults can exceed 71 inches in carapace length and weigh more than 1,100 pounds. The gray to black carapace lacks the scutes of other marine turtles and is instead leathery or rubberlike with numerous white flecks or spots. Five distinct ridges run parallel on the carapace from head to tail.

An adult female leatherback nesting.

Note to Readers: For detailed descriptions of the appearance of hatchlings and adults, habitat preferences, behavior, life cycles, and conservation issues related to sea turtles of the Southeast, see *Sea Turtles of the Atlantic and Gulf Coasts of the United States* (University of Georgia Press, 2006) by Carol Ruckdeschel and C. Robert Shoop.

U.S. SPECIES NOT FOUND IN THE SOUTHEAST

Forty-two of the 56 turtle species known from the United States are found in the Southeast. The rich biodiversity of turtles here is readily demonstrated by the U.S. map showing the high concentrations of species in the Gulf Coastal Plain (see map on page 236). The following cameo accounts of the 14 turtle species that are not inhabitants of the Southeast underscore the significance of southeastern wetlands and the overall habitat diversity of the region for turtles. Of the native U.S. species of turtles found outside the Southeast, 2 are terrestrial tortoises, 6 are semiaquatic, 5 are riverine, and 1 is a Pacific marine species. Five of the 14 species (see below) are found in eastern Texas, adjacent to the southeastern states.

Did you know?

The Southeast has the highest diversity of turtles in the United States and is home to 75 percent of all native species found in the United States and Canada.

Five of the 14 native U.S. species found outside the Southeast are native to eastern Texas, adjacent to the southeastern states: the Texas cooter (top left), Cagle's map turtle (top right), the Texas map turtle (bottom left), the Texas tortoise (bottom center), and the yellow mud turtle (bottom right).

Desert Tortoise *Gopherus agassizii*

The desert tortoise, a close relative of our gopher tortoise, occurs in the Mojave and Sonoran deserts of the Southwest. Its habitat is both drier and hotter in the summer and colder in the winter than that of the gopher tortoise. The desert tortoise is declining due to a contagious respiratory disease, drought, and human development.

Texas Tortoise *Gopherus berlandieri*

The Texas tortoise occurs in southern Texas and parts of Mexico in scrub and desertlike areas. Less is known about this North American tortoise than the other two tortoise species. It does not dig an extensive burrow like the gopher tortoise but instead makes a shallow resting place by nestling down under a bush or cactus. The Texas tortoise is declining as a result of habitat alterations and road mortality.

Although not actually recognized as a species separate from the painted turtles of the Southeast, western painted turtles (*Chrysemys picta bellii*) are uniquely colored and patterned to the extent that they might appear to be different species.

Arizona Mud Turtle *Kinosternon arizonense*

The Arizona mud turtle, a species of the Sonoran Desert in Arizona and Mexico, lives in temporary ponds and springs on the desert floor. It is inactive during dry periods, which is much of the time in that region, and emerges during summer rains when the ponds fill. Its general behavior is similar to that of mud turtles in the Southeast, except that its periods of activity are much more limited due to the frequent lack of standing water.

Yellow Mud Turtle *Kinosternon flavescens*

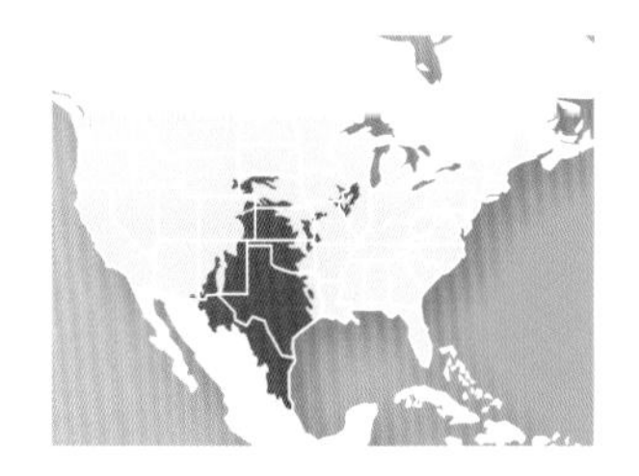

The yellow mud turtle is primarily a species of the Great Plains. Although it is found in nearby eastern Texas and southwestern Missouri, it has never been reported in adjacent Louisiana and Arkansas. The yellow mud turtle can be found in any type of aquatic habitat, although seasonal ponds are favored. It hibernates on land and digs deep below the frost line. It often moves overland between ponds during warm weather.

Big Bend Mud Turtle *Kinosternon hirtipes*

The Big Bend mud turtle is primarily a Mexican species and is known in the United States only from the Big Bend region of Texas. It resembles the Southeast's eastern mud turtle but is more elongated. The Big Bend mud turtle has been found in a few permanent springs in this desert grassland region and likely occurs in the Rio Grande.

Sonoran Mud Turtle *Kinosternon sonoriense*

The Sonoran mud turtle, which is distributed widely throughout the Sonoran and Chihuahuan deserts of the southwestern United States and Mexico, is larger than either of the southeastern mud turtles. It is associated with permanent and intermittent streams and springs within the deserts where it lives. Increasing drought severity has caused many once-permanent desert streams to become seasonal in their flow, perhaps having severe environmental impacts on this species.

Pacific Pond Turtle *Clemmys marmorata*

The Pacific pond turtle is the only semiaquatic turtle native to the West Coast. This medium-sized turtle inhabits streams, slow-moving sections of rivers, ponds, and freshwater marshes and can often be observed basking on rocks and logs. Loss of habitat and a mysterious lack of hatchling recruitment have been cited as causes of its decline. Its closest relatives are probably the spotted, bog, and wood turtles of the Southeast, as well as Blanding's turtle.

Blanding's Turtle *Emydoidea blandingii*

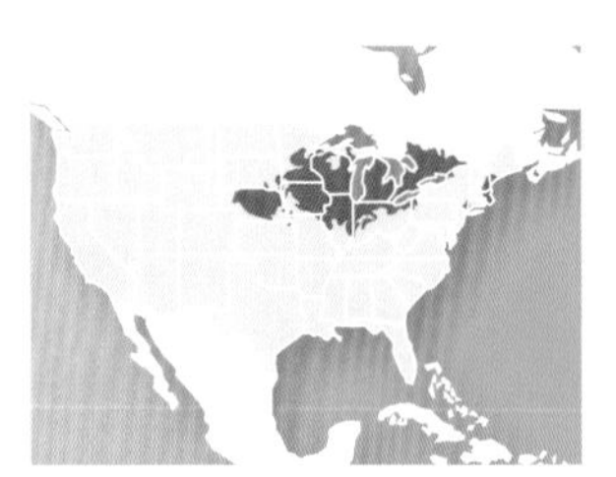

Blanding's turtle is a large semiaquatic turtle of the northern United States. Nearly all of its current range lies within areas that were once glaciated, although fossil remains of Blanding's turtles have been found in South Carolina. This turtle inhabits grassy freshwater marshes, duckweed-covered swamps, and seasonal wetlands where it feeds primarily on crayfish and tadpoles. Its bright yellow throat makes it easy to identify as it sticks its head above the duckweed or water lily–covered marsh surface. Because this species frequently travels overland between different ponds and marshes, it is susceptible to highway mortality.

Big Bend Slider Turtle *Trachemys gaigeae*

The Big Bend slider is obviously different from the sliders of the Southeast. A round orange spot marks each side of the head behind the eye. The yellow plastron is conspicuously marked with a very colorful black-and-orange pattern along the seams. This slider is closely associated with the Rio Grande in New Mexico and Texas. Its historical habitat was floodplain oxbows and freshwater marshes, but channelization and removal of water for irrigation have significantly reduced the amount of habitat available to this turtle. It continues to persist in surrogate habitat in several New Mexico reservoirs and riverine pools, but it may be one of the most threatened turtles in the United States.

Rio Grande Cooter *Pseudemys gorzugi*

The Rio Grande cooter is a large, colorful relative of the Southeast's river cooter. Reds, pinks, and oranges color the plastron and the limbs. As its common name implies, this cooter lives in the Rio Grande on the U.S.-Mexico border, as well as in the Pecos River in southeastern New Mexico. It prefers permanent flowing streams and large rivers that contain deep pools and clear water; it is absent from muddy, shallow, and sediment-laden sections. Erosion from surrounding agricultural and grazing lands as well as ever more frequent droughts threaten this species.

Texas Cooter

Pseudemys texana

The Texas cooter is a little-known riverine species that occurs in central and south Texas in the Colorado, Brazos, Guadalupe, and San Antonio river systems. Its yellow head striping differs from that of river cooters of the Southeast in that the lower stripe curves around the angle of the jaws, and a small yellow spot occurs behind each eye. The Texas cooter needs further study to determine its ecology and conservation needs.

Cagle's Map Turtle

Graptemys caglei

Cagle's map turtles are found in the San Antonio and Guadalupe rivers of southeastern Texas. They are similar in general appearance to Ouachita map turtles but are distinguished by a transverse yellow bar under the chin. They live in streams that originate in karst (limestone rock) landscapes and prefer pools with moderate flow rates. Males pluck and eat aquatic insects from sunken log snags; the slightly larger females eat snails from the stream bottoms. They are declining as a result of dams and collection for the pet trade.

Texas Map Turtle

Graptemys versa

Texas map turtles are endemic to the Colorado River system of central Texas (not the Colorado River of Grand Canyon fame) within the Edwards Plateau. They live in clear streams that originate in this limestone-dominated region. They are more omnivorous than other map turtles, consuming some aquatic plants as well as insects. Adults are relatively small, although females are larger than males. Little is known about their conservation needs.

Olive Ridley Sea Turtle *Lepidochelys olivacea*

The olive ridley sea turtle is a tropical ocean species known from the Pacific and Indian Oceans as well as the southern Atlantic. Individuals are occasionally spotted in the Caribbean. Most U.S. sightings are along the California and Oregon coasts, where these sea turtles forage on snails, jellyfish, crabs, and other crustaceans during warm summers. They nest on Pacific beaches of Costa Rica, Mexico, and South America. Harvest of adults and eggs for human consumption around the world has led to overall population declines.

A herpetologist uses calipers to track a hatchling's growth.

People and Turtles

WHAT IS A HERPETOLOGIST?

Herpetologists are scientists who study amphibians and reptiles. Frogs and salamanders are amphibians, and snakes, lizards, and turtles are reptiles; collectively, amphibians and reptiles are called herpetofauna. Herpetologists usually specialize in a particular biological discipline, such as ecology, behavior, or genetics. Because so many species of amphibians and reptiles throughout the world are declining in numbers, many herpetologists devote their scientific efforts toward conservation issues—for example, investigating the impacts that draining or filling wetlands can have on amphibian and reptile populations. Many specialize on particular groups, such as frogs or turtles, or even on particular species.

Herpetologists have entered the field from many different paths. Some got hooked as children when they found a painted turtle in a nearby pond or rescued a box turtle crossing the road. Others had a pet turtle bought in a pet shop. Some got interested in herpetology from a high school or college biology class field trip. Still others started studying herpetofauna later in life, perhaps after beginning a career as a wildlife biologist or manager. Many scientific papers are written each year on amphibians and reptiles, and some scientific journals publish only herpetological research, al-

Ecological studies monitor and evaluate the status of turtles in particular habitats.

A hoop net with guide nets attached can be used to intercept and capture turtles.

though the research is often applicable to other animals besides amphibians or reptiles.

Why Do Herpetologists Study Turtles?

Herpetologists study turtles for many reasons. Turtles are particularly interesting to ecologists because their life spans are so much longer than those of other reptiles. Some paleontologists have focused on turtles because of the excellent and extensive fossil record that exists for them, particularly in comparison with snakes and lizards. Many conservation biologists concentrate their field research on species or groups of turtles because of the many environmental threats these animals face. For example, desert tortoises in the Southwest and gopher tortoises in the Southeast have been extensively studied because both are so susceptible to human development encroaching on their natural habitats. Research on the basic ecology, behavior, and causes of mortality of threatened or endangered species is necessary to understand how to preserve them and their habitats. Ecological studies that monitor and evaluate the status of turtles in particular habitats or environmental situations can lend insights that help conservation biologists address active or potential environmental problems.

A frame with netting can capture turtles as they fall from favorite basking logs.

How Do Herpetologists Study Turtles?

Herpetologists use a variety of techniques to study turtles, many of them especially designed for turtles' unusual body structure and behaviors. Among the commonly used techniques for capturing turtles are hoop traps that are partially submerged in aquatic habitats. The traps can be baited or have fencelike extensions that lead turtles into the mouth of the funnel. Basking traps can be highly effective in some situations. A basic basking trap consists of a square frame with floats on opposite sides and a mesh bottom. In one design, wooden boards allow turtles to climb up out of the water and bask, but when they drop back into the water they are inside the frame. Hoop traps and basking traps have been modified in many ways depending on the circumstances and the species to be captured, but the basic principles are the same: the turtle enters an enclosed area and can be collected by the turtle biologist.

Both aquatic and terrestrial turtles can be captured as they cross roads in the vicinity of the study area. Unfortunately, many turtles are killed by automobiles, but even dead specimens can be used to provide information

Snorkeling for alligator snappers (above) can be an exhilarating experience.

Dipnetting turtles from a canoe (left) requires skill and teamwork.

Drift fences with pitfall traps can be used to intercept turtles as they migrate between aquatic and terrestrial habitats.

about the species' biology, such as number of eggs a female was carrying or what food was in the stomach. One highly effective means of capturing turtles systematically is to use a drift fence around an aquatic habitat. A fence of wire mesh, aluminum siding, or netting is erected in an area where turtles are likely to leave or enter the water. Buckets are sunk in the ground along both sides of the fence at regular intervals. A turtle that encounters the drift fence tends to turn in one direction or the other and follow it, and eventually falls into an open bucket trap. The drift fence technique allows the investigator to determine the species of turtles leaving or entering, when they are moving, and which direction they are going. Many other specialty techniques have been used to capture turtles, including seining, gill and trammel netting, snorkeling, and feeling, or "muddling," for aquatic or semiaquatic species in the mud or under the banks of aquatic areas. Dogs have even been trained to locate box turtles by their scent.

Did you know?

Like bones, turtle egg shells contain calcium. X-rays can be used to determine how many eggs a female is carrying.

In ecological studies with turtles, several standard measurements are made on captured individuals, including plastron and carapace length and body weight. Determining the sex of the animal is easy to do with adult turtles of most species because the males have longer tails than the females. Unusual color patterns or injuries, such as a broken shell or missing limbs, are also recorded. Some turtle biologists take blood or muscle tissue samples for DNA or health analyses. Many turtle biologists use the mark-release-recapture approach to assess population size, to determine growth rates of individuals, and to estimate mortality rates in the population. The method most commonly used to mark hard-shelled turtles is to give each marginal scute a letter or number and to notch or drill holes in these scutes, using different combinations of notches to give each individual a unique code.

Radiotelemetry has been used to study the natural behavior of turtles in both terrestrial and aquatic habitats. A radiotransmitter attached with epoxy to the shell of a turtle emits a signal at a particular frequency at

MARKING TURTLES FOR LONG-TERM STUDIES IN THE FIELD

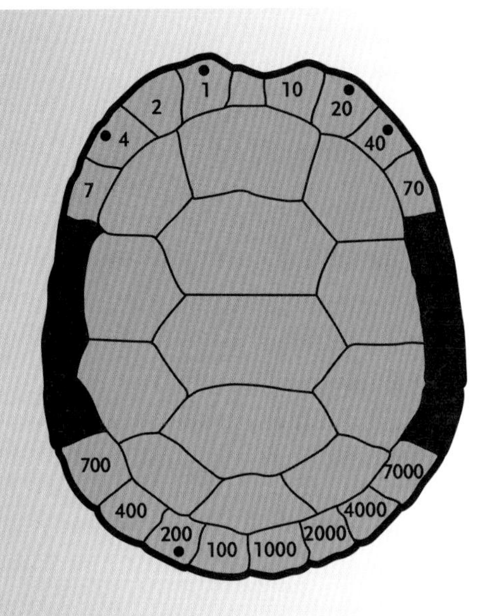

A marking system using numbers for turtles. The 1-2-4-7 system allows for any number between 1 and 9 by marking or drilling (illustrated here). The identification code for this turtle is 265. Up to 9999 different turtles can be identified with this method.

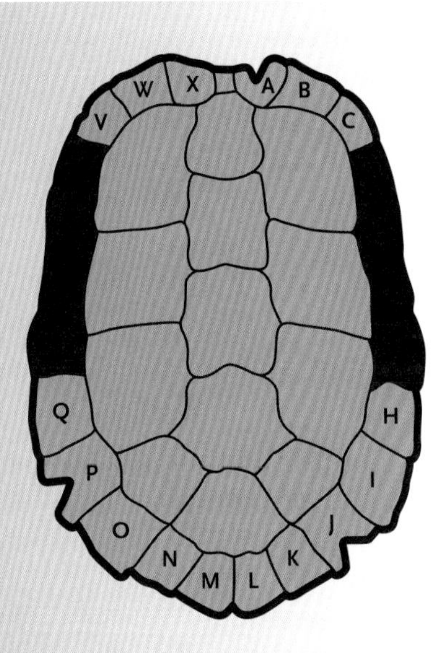

A marking system using letters for slider turtles. The marginal scutes indicated by shading should not be included in the marking scheme for certain species because of their susceptibility to shell damage. The identification code for this turtle is AJP.

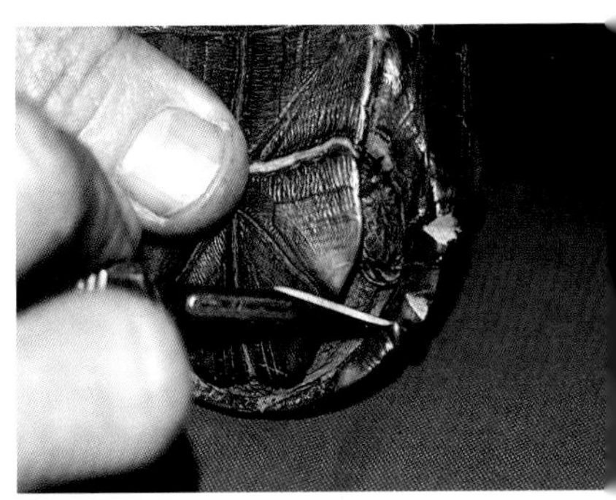

The marks on turtle shells can be made as notches from a knife or file (bottom right) or as holes drilled through the marginals (above left). Hatchlings can be marked with small scissors or fingernail clippers (top right). Marks on the shells of terrestrial and freshwater turtles do not damage the turtle but are permanent and can be identified as many as 30 years after initial marking.

Radio transmitters can be attached with epoxy.

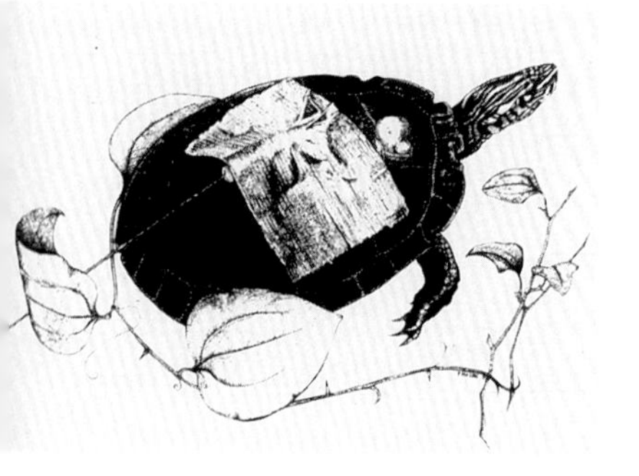

A spool of thread attached with duct tape is a simple way to determine where a turtle goes. *Artwork by Peri Mason.*

regular intervals. Thus, the researcher can use a receiver and a directional antenna to determine the location of the turtle at any time. Radiotracking can be used to study behavior, movement patterns, and habitat selection. X-rays have proved invaluable in the study of turtle reproduction all over the world because they permit a harmless assessment of the number of shelled eggs a female is ready to lay.

What Rules Must Herpetologists Follow to Study Turtles?

The collection of turtles for research or educational purposes—including for a college class in herpetology or a research project approved by a student's academic committee—may still be affected by state and federal laws that protect turtles and other wildlife. It is the turtle researcher's responsibility to be aware of and to follow all state and federal regulations. In the southeastern states, anyone removing turtles from the wild for research purposes, even if the turtles are to be released later, must have a collecting permit. Some states place limits on the number of turtles of some species that can be held at one time. Southeastern turtles protected under the federal Endangered Species Act, such as the Alabama red-bellied cooter and flattened musk turtle, also require federal permits before these animals can be handled in the field or studied in the laboratory. Learning the laws of each state where research is being conducted, doing the paperwork to obtain permits, and paying a fee can sometimes seem onerous, but most wildlife rules and regulations designed to protect turtles are in the overall best interest of the turtle populations. Conversely, the data turtle researchers provide to resource management agencies should help determine future conservation actions that benefit turtles.

KEEPING TURTLES AS PETS

Turtles were common pets in the 1940s and 1950s, when dime stores sold baby turtles by the thousands. In the 1970s, the U.S. Food and Drug Administration made the commercial sale of turtles less than 4 inches long illegal. The rationale was that baby turtles were a common source of the *Salmonella* bacterium, which can cause a type of food poisoning in children that have handled infected turtles.

Although federal size regulations restrict the sale of baby turtles by pet stores and reptile breeders, keeping captive turtles is generally not difficult if guidelines of health safety, for the keeper and the turtle, are followed. Cleanliness is essential. Small water bowls provided for terrestrial turtles need to be changed daily, and larger containers for aquatic turtles should be changed frequently enough to keep the water clear and without visible food particles. Circulating water systems with filters are the most satisfactory approach for keeping aquatic turtles in a clean habitat. The presence of *Salmonella* is usually a consequence of infrequent or improper cleaning of cages and water bowls and can be avoided with these routine procedures. Always wash your hands after handling turtles. Some people keep terrestrial box turtles outdoors in an enclosed backyard or pen.

Many books are available on the care and maintenance of pet turtles, and it is important to consult one that outlines the particular needs of the species you plan to keep. Basking sites and occasional exposure to natural

Most children are captivated by turtles.

Well-designed backyard turtle ponds and gardens provide enjoyment for all members of the family.

sunlight are important for aquatic species such as painted turtles and sliders, for example, while box turtles in an outside enclosure will require shaded areas during summer. Food requirements also vary with the species; some are mostly herbivorous while others are primarily carnivorous. Proper care also includes being aware of species that do not do well in captivity.

How Do You Get a Pet Turtle?

While catching it yourself is one of the most personally gratifying means of acquiring a turtle for a pet, herpetologists and conservation biologists disagree on the advisability of removing an animal from its natural habitat. One way to obtain a pet turtle that does not interfere with nature is to get a captive-bred individual through a local or regional herpetological society. Several herpetology clubs provide turtles for adoption to their members. These turtles are often from unknown origins and cannot be released back into the wild. Many legitimate pet dealers sell captive-bred individuals.

Whatever the best approach is from a conservation standpoint, it is certain that most professional herpetologists and reptile conservation biologists kept snakes, turtles, or other wild animals in captivity as children and learned from their experiences. Children who learn to appreciate nature are more likely to become conservation-minded adults. We believe that children must be able to go out into a pond or woods and handle a turtle, frog, or snake. We do not suggest that a wild turtle should be brought home and kept permanently as a pet.

Some turtles can bite with authority, so catching or keeping a turtle like a large softshell or snapping turtle requires caution. Catching a baby is always the safest approach. One of the most enjoyable means of obtaining a pet turtle is to hatch it yourself from an egg. Eggs can sometimes be obtained from road-killed females or from nests that have been partially destroyed by predators. Fresh eggs are often still viable and can be incubated until they hatch.

The simplest way to incubate turtle eggs is to place them between paper towels that are kept slightly damp, or to partially bury them in vermiculite or the soil in which they were originally deposited. Depending on the species, of course, the eggs normally take 6–10 weeks to hatch at 75–80°F. Do not rotate incubating turtle eggs. Unlike bird eggs, which the mother turns periodically in the nest, turtle eggs must keep the same orientation once incubation begins because the turtle embryo develops on top of the yolk at the highest point within the egg. If the egg is rotated, the embryo will smother and die. After hatching, the baby turtle may require several days to absorb its delicate external yolk sac.

Can I Keep It?

One issue that must be considered with all groups of wild animals is whether laws and regulations govern their capture and captivity. Such laws may be federal and apply wherever the species occurs, or state or local and apply to turtles found within that region. If a turtle species is on the federal endangered species list, a live one cannot be picked up, and the eggs, shell, or other body parts cannot be kept without a permit. The degree of protection turtles and other reptiles receive varies considerably among the southeastern states, and those planning to keep a pet turtle (whether wild caught or captive raised) should become familiar with the regulations of their state's wildlife department. The sale of turtles as pets is governed by additional laws that can include city, state, federal, and international jurisdictions.

Practical and Ethical Issues

Anyone keeping a wild animal as a pet should consider certain traits of the species that could have ultimate effects on the environment, wild populations of the species, and the individual animal itself. One issue with turtles is that individuals live a long time compared with other animals. Are you prepared to keep your baby painted turtle for 30 years? Even small mud turtles may live to be 40, and box turtles can live more than 50 years. Keeping a turtle could be a lifetime commitment, so be prepared.

Turtles not only live a long time, some grow to large sizes that make them difficult to keep in a home aquarium or terrarium. What do you do when the inch-long slider turtle you found in a local lake reaches several inches in length? Returning it to the lake is one obvious solution, but certain issues must be considered. Will the turtle be able to feed itself after you have hand-fed it for several months or years? Could it have contracted a disease that can be transmitted to the wild population? Are you able to return it to the same place where you caught it? People move, and habitats change or become inaccessible. Releasing a turtle outside its natural range or into an inappropriate habitat is not only bad for the turtle; it may not be good for the local environment. When turtles are released into areas outside their natural range, genetic mixing may occur. The extent or impact of genetic alterations within turtle populations is not easily assessed, but the natural genetic makeup of turtles in the area is clearly affected.

Did you know?

Red-eared sliders have become established all over the world and may be harmful to native species if they outcompete them for resources.

Kids view turtles from the boardwalk at Riverbanks Zoo.

Turtle Educational Exhibits

Turtles of the Southeast are popular subjects for regional zoos, aquariums, and nature reserves. Some of the best exhibits have large aquatic areas indoors where turtles can be viewed as they swim underwater and outdoor boardwalks where turtles can be seen basking. Such exhibits provide an opportunity for people to experience and appreciate species that they might not encounter in the wild.

TURTLES AS FOOD

In most regions of the world, turtles are distinct from most other reptiles because their meat and eggs are considered a source of food. Several species of southeastern turtles are considered delectable, and almost any species can be eaten. Gopher tortoises were once called "Hoover chickens" because they were eaten by rural southerners during the Great Depression when Herbert Hoover was president. Sliders, cooters, and chicken turtles all have

Turtles have been exploited commercially for food and other products.

Author Kurt Buhlmann examining turtles for sale in a market in China.

edible meat on the limbs, plastron, and carapace, but the process of catching, killing, and cleaning them is hardly worth the effort. The diamondback terrapin, however, an even smaller turtle, was once much sought after for its meat. During the early part of the twentieth century this species of the coastal salt marshes was heralded along the Atlantic Coast as a reptile delicacy unsurpassed in flavor. Terrapin soup became so popular that diamondback terrapins became one of the earliest examples of a commercially extinct species; that is, their numbers dwindled so low that the cost and effort of capturing them was no longer profitable and the terrapin soup industry declined.

Sea turtles historically have been the most popular food, especially the green sea turtle, which is highly valued for its meat and eggs. North Americans ate sea turtle meat and eggs when they were available, but farther south in tropical America, where green sea turtles nest in even greater numbers, they became a dietary staple. The protection given all U.S. sea turtles under the Endangered Species Act precludes further consumption within the United States, but sustenance consumption and commercial exploitation for food still occur in many other regions of the world.

During the past half century, commercially supplied turtle meat in the United States has come primarily from large freshwater species—softshell, common snapping, and alligator snapping turtles. Commercial sales of turtles for meat in the United States have diminished since the late twentieth century, in part because of more stringent state regulations and in part because of changing public attitudes about native wildlife. American turtle meat is still exported to other parts of the world, however, especially red-eared sliders, alligator snappers, and Florida softshells, although records of the exact amounts exported are difficult to obtain. The true toll that early commercial sales for the food industry had on wild populations of turtles will likely never be known, but current export levels should be evaluated to ensure that southeastern turtle populations do not decline.

Commercial turtle farming is big business in China (top).

In Vietnam, local residents raise turtles for food in backyard ponds (middle).

Aquatic turtles are sometimes raised in captivity in large tanks (bottom).

DANGER
Do Not Lift Children to
Top of Viewing Window

Passageways under roads that permit movement of turtles and other animals must be included in highway construction designs.

Ensuring the persistence of some species, such as these gopher tortoises, may require reintroductions.

A Conservation Vision for Turtles

Turtles have lived on the earth since before the dinosaurs. The group has survived changing climates and escaped many types of predators. However, studies that began during the latter half of the twentieth century and continued into the twenty-first indicate that most of the world's tortoises, freshwater turtles, and marine turtles are becoming increasingly imperiled.

Although the purpose of this book is not to solve the problems facing turtles around the world, understanding the reasons for their decline may help to change attitudes and lead to solutions. Our overriding purpose in writing this book is to foster appreciation for turtles and tortoises, animals we have spent our professional lives learning about and are now involved in helping to conserve. It is clear that people will protect only what they understand and care about.

Recognized reasons for population declines of reptiles globally include (1) habitat loss and degradation, (2) introductions of invasive species, (3) environmental pollution, (4) diseases and parasites, (5) unsustainable use, and (6) global climate change. Many of these problems are present in the Southeast.

Children who learn about natural history are more likely to understand the needs of the environment.

The rural Southeast is changing. Urban development and sprawling subdivisions have eaten up farm and timber land. Human populations are growing and require more roads, water, and other services. The loss of habitat has become an overriding problem for all native wildlife in the region, including turtles. Knowledge about where certain turtle species occur on the landscape and which habitats they must have for their continued existence is the first requirement for productive discussions with land managers, such as foresters and farmers, as well as local planners and developers.

Although we know a great deal about the habitats and habits of certain turtles, there is still much to learn. Questions about what turtles do and how they live in their environment are first asked by young children who play and explore outdoors. Their insatiable curiosity often develops into an interest in natural history. That interest must be fostered and nurtured in future generations. Not all children need to become professional ecologists or herpetologists, but children who learn about natural history and grow up to be developers or businesspeople are more likely to appreciate the need to balance growth and development with the needs of the environment. Children who do not connect with nature on some level are less likely to see how the natural environment, including turtles, is important to their lives. Education about the functions of the environment and how our actions affect it is valuable for all citizens. There are too many examples of a habitat needed by a particular species being destroyed or polluted by inadequate planning and development.

Locks and dams block the movements of turtles, as well as fish and other creatures.

Active habitat management, including controlled burning, is needed to maintain or restore some turtle habitats.

What specific actions can citizens of the Southeast take to prevent declines in native turtle populations? What actions can we take to recover and restore populations and their habitats? Efforts to restore habitat for certain species, including turtles, are currently under way. Private landowners, natural resource managers, and state, federal, and university scientists are working together to make a positive difference. Private individuals and zoos that breed rare species may be able to contribute animals to reintroduction efforts in conjunction with habitat restoration efforts.

Young children must be allowed to experience turtles in the wild, or else they may not care about them as adults.

Those of us living in the biodiversity-rich Southeast need to know what we have and where it lives, and then we must have the appreciation and vision to protect it. Wetland habitats, especially small, isolated ones, have minimal protection under state and federal laws. But as this book has made clear, these wetlands are home to many of the Southeast's turtles. Some turtles also need to travel safely across the landscape between wetlands. Landowners, town planners, and concerned citizens who understand the value of these places can work together to either protect them or minimize impacts on them. Perhaps buffers of forested habitat can be preserved around a wetland; perhaps the path of a road can be replanned to avoid severing the connection between two wetlands; perhaps a road sign can be installed that warns motorists that turtles frequently cross at that point.

Upland habitats are also threatened. Gopher tortoises are losing the sandy habitats they need as the human population of Florida and other southeastern states grows. Clearly, large tracts of land must be set aside for gopher tortoises and the other unique plants and animals that share the sandy uplands. Picking up tortoises and moving them out of harm's way is admirable, but at best that will save only individuals, not populations. These social, long-lived animals need sites with long-term stability, management, and protection if they are to persist.

River habitats in the Southeast are in need of help, too, and addressing the numerous issues they face is an enormous challenge. Rivers receive many pollutants from the landscape, and the sources are difficult to trace. Fortunately, the citizens of the Southeast value the region's rivers for recreation and commerce, and the actions they take to allow rivers to function normally will also help maintain the biodiversity of rivers, including the turtles.

Our wetlands, uplands, and rivers can be better protected and managed, and many restoration projects are already under way. Minimizing the loss of additional important habitats is also necessary for the long-term well-being of the landscapes of the Southeast, its plants and animals, and its human citizens.

What kinds of turtles are found in your state?

Occurrence by state of the 42 species of turtles native to the Southeast, listed in the order in which they appear in the book

COMMON NAME	LA	AR	MS	AL	GA	FL	SC	NC	VA	TN	KY
TERRESTRIAL TURTLES											
Gopher tortoise	●		●	●	●	●	●				
Ornate box turtle	●	●									
Common box turtle	●	●	●	●	●	●	●	●	●	●	●
SEMIAQUATIC TURTLES											
Wood turtle									●		
Bog turtle					●		●	●	●	●	
Spotted turtle					●	●	●	●	●		
Common mud turtle	●	●	●	●	●	●	●	●	●	●	●
Striped mud turtle					●	●	●	●	●		
Chicken turtle	●	●	●	●	●	●	●	●	●		
Painted turtle	●	●	●	●	●		●	●	●	●	●
Slider turtle	●	●	●	●	●	●	●	●	●	●	●
Pond cooter	●		●	●	●	●	●	●	●		
Florida red-bellied cooter					●	●					
Northern red-bellied cooter								●	●		
Common snapping turtle	●	●	●	●	●	●	●	●	●	●	●
Common musk turtle	●	●	●	●	●	●	●	●	●	●	●
Florida softshell turtle				●	●	●	●				
RIVERINE TURTLES											
Alligator snapping turtle	●	●	●	●	●	●				●	●
Razor-back musk turtle	●	●	●	●							
Loggerhead/Stripe-necked musk	●		●	●	●	●		●	●	●	
Flattened musk turtle				●							
Spiny softshell turtle	●	●	●	●	●	●	●	●	●	●	●
Smooth softshell turtle	●	●	●	●		●				●	●
Alabama red-bellied cooter			●	●							
River cooter	●	●	●	●	●	●	●	●	●	●	●
Barbour's map turtle				●	●	●					
Alabama map turtle			●	●	●						
Ernst's map turtle				●		●					
Gibbons' map turtle	●		●								
Common map turtle		●	●	●	●				●	●	●
False map turtle	●	●	●							●	●
Ouachita map turtle	●	●	●	●						●	●
Sabine map turtle	●										
Black-knobbed map turtle			●	●							
Yellow-blotched map turtle			●								
Ringed map turtle	●		●								
BRACKISH WATER TURTLES											
Diamondback terrapin	●		●	●	●	●	●	●	●		
MARINE TURTLES											
Loggerhead sea turtle	●		●	●	●	●	●	●	●		
Kemp's ridley sea turtle	●		●	●	●	●	●	●	●		
Green sea turtle	●		●	●	●	●	●	●	●		
Hawksbill sea turtle	●		●	●	●	●	●	●	●		
Leatherback sea turtle	●		●	●	●	●	●	●	●		
TOTAL	27	16	30	30	27	25	21	21	23	15	13

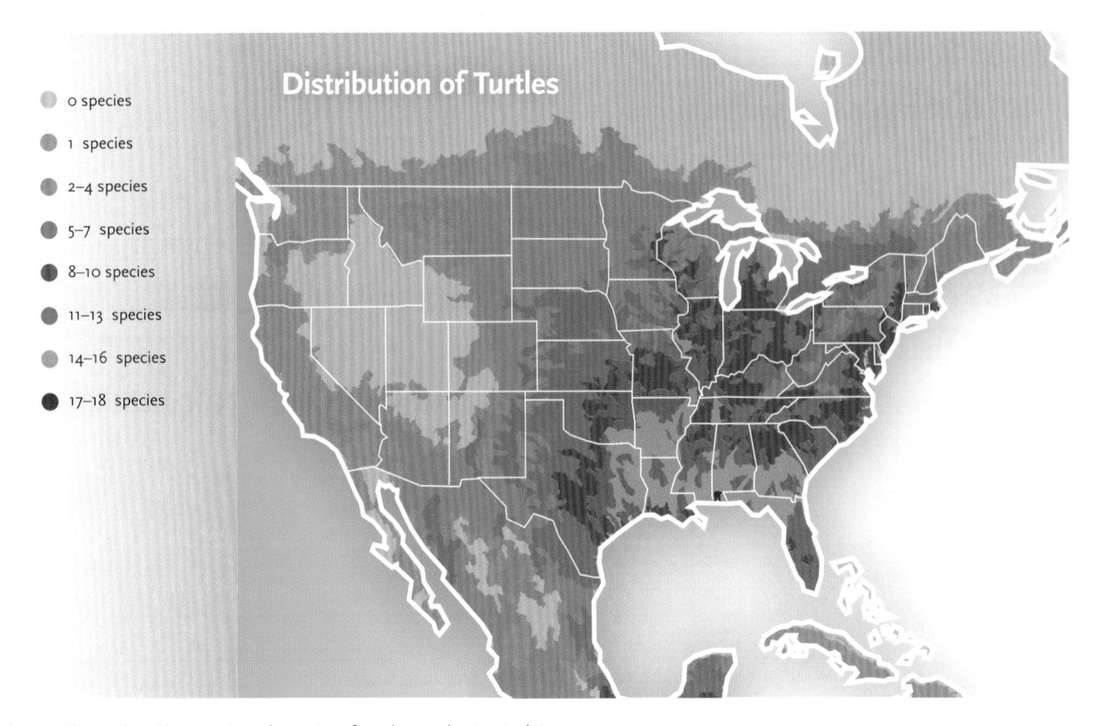

The Southeast has the greatest diversity of turtles in the United States.

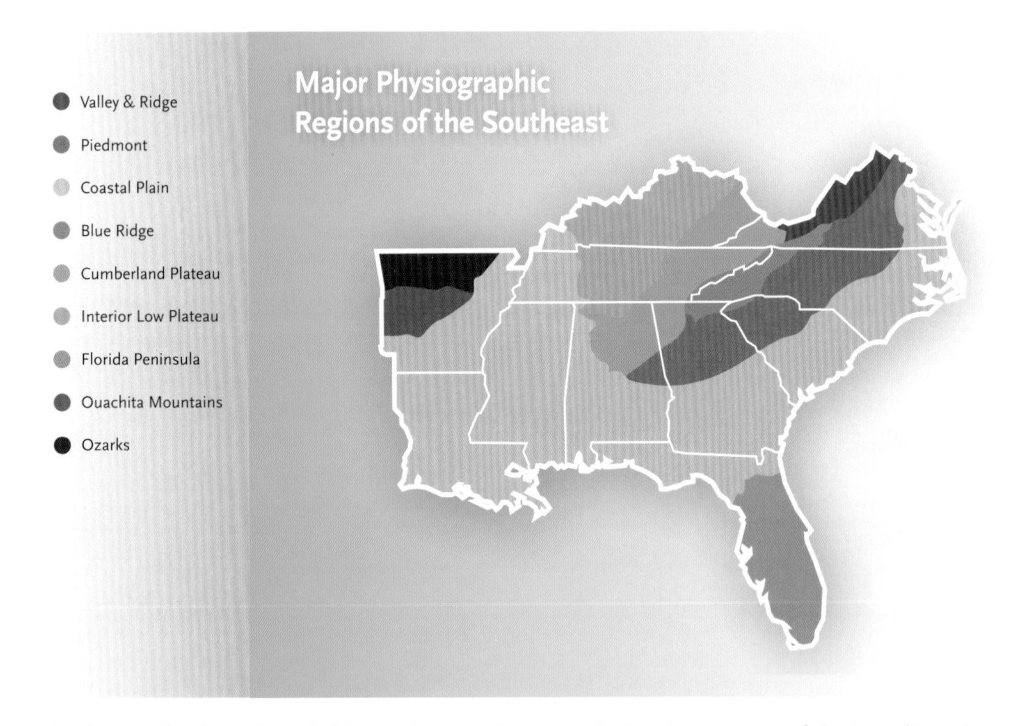

The distribution of turtles and their habitats is determined in part by the broader categories of physiographic regions.

Glossary

Aestivation A period of inactivity during dry and/or hot periods when animals wait for conditions suitable for feeding and other activities to improve.

Albino An animal completely lacking pigment that provides the color to skin and eyes; animals lacking only dark pigment, or melanin, are often referred to as "amelanistic."

Annuli (*sing.* annulus) Growth rings on the scutes of turtles that represent periods of rapid growth (such as during the warm period of the year) and slow growth (such as during winter); many southeastern turtles produce annuli yearly.

Anterior Referring to the end of the animal toward the head.

Barbel A fleshy projection on the chin or neck of some species of turtles that may be used to locate food by touch at night or in turbid water.

Bask To expose the body to sunlight either out of the water, as on logs or rocks, or while at the water's surface, such as by a turtle resting on vegetation.

Biodiversity Referring to the numbers, distribution, and abundance of species within a given area.

Bioindicator A species whose health or condition, either at the individual level or at the population level, indicates the condition of the habitat or ecosystem as a whole.

Brumation A period of inactivity by "cold-blooded" animals, or ectotherms, during cold periods. *See also* Hibernation.

Carapace The upper shell of a turtle.

Carolina bays Isolated, egg-shaped wetland habitats, variable in size, that are common to the Coastal Plain from Georgia through the Carolinas and are home to several species of turtles.

Caruncle Egg tooth; a temporary, pointed structure that develops on the end of the snout of turtles prior to birth and is used to slice open the eggshell.

Cloaca A single opening that serves as the passageway to the outside of an animal for the urinary, digestive, and reproductive tracts.

Clutch A group of eggs laid together at one time by a single individual.

Cold-blooded A nontechnical term that refers to animals whose body temperature is largely determined by environmental conditions and the thermoregulatory behavior of the animal. *See also* Ectotherm.

Costal scutes The lateral row of scutes on the carapace of a hard-shelled turtle that run lengthwise down the carapace on either side of the vertebral scutes; costal scutes are also called pleurals or laterals.

Courtship Behavior designed to increase a female's willingness to mate; the male may bump or bite her shell, titillate her with his long foreclaws, or use other behavior to make her more receptive. *See also* Titillation.

Cusps Pointed, bony projections on the upper jaw of red-bellied cooters presumably used for shredding vegetation and useful in distinguishing these species from other cooters.

Diurnal Active during the daytime.

Dorsal Referring to the top of a turtle.

Ecology The study of how organisms interact with their environment.

Ectotherm An animal whose body temperature is largely determined by environmental conditions and the thermoregulatory behavior of the animal. *See also* Cold-blooded.

Endangered Referring to a species or population that is considered at risk of becoming extinct.

Endemic Referring to a species found only in a particular geographic location and nowhere else.

Endotherm An animal that maintains a high body temperature primarily through the use of heat generated by a high metabolic rate. *See also* Warm-blooded.

Extinct Referring to species with no living individuals.

Extirpated Referring to the elimination of a species or population from a particular region; local extinction.

Fall Line The physiographic boundary separating the Piedmont from the Coastal Plain, where rivers typically change from fast flowing and rocky to meandering and sandy bottomed.

Family A taxonomic group containing one or more closely related genera.

Foreclaws The claws on the front feet of turtles.

Form A shallow hideaway created beneath leaves or other ground litter by box turtles and by some semiaquatic turtles when on land.

Generalist An animal that does not specialize on any particular type of prey or is not restricted to a particular habitat.

Genus (*pl.* genera) A taxonomic grouping of one or more than one closely related species.

Gular One of the pair of most anterior plates on the plastron of a turtle.

Herpetofauna The amphibians and reptiles that inhabit a given area.

Herpetologist A scientist who studies turtles, other reptiles, and amphibians.

Hibernation A period of inactivity during cold periods; also known as "brumation" in reptiles.

Hinge Flexible section across the plastron of some turtles that allows the lower shell to touch the upper shell (carapace) so that the soft parts of the body are enclosed within the shell.

Hybridization Mating between two different species that results in offspring.

Incubation period The time period between when eggs are laid and when they hatch.

Intergrade (Intergradation) An intermediate form of a species resulting from mating and genetic mixing between individuals of two or more subspecies within a zone where their ranges overlap; intergrade specimens may possess traits of all subspecies involved.

Keeled carapace A carapace that has a ridge down the center running from front to back.

Lingual lure The fleshy, pinkish, tonguelike structure visible in the open mouth of an alligator snapping turtle that is used to attract prey.

Marginals The scutes, usually numbering 22 or 24, that form the edge of a turtle's carapace.

Melanism *(adj.* melanistic*)* A condition especially characteristic of adult male slider turtles in which the amount of black pigment (melanin) in the shell and skin is increased and masks other colors that may have once been visible.

Nocturnal Active at night.

Nuchal The anteriormost scute of a turtle carapace at the point where the marginal scutes from the two sides meet.

Ocellus *(pl.* ocelli*)* A colorful eye-shaped marking on the shell of some turtles.

Omnivores Animals that eat a variety of plant and animal foods.

Oxbow A U-shaped bend of a former river channel typical of large, southeastern rivers in the Coastal Plain below the Fall Line.

Phylogeny The evolutionary relationships among different groups and species of animals.

Physiographic Referring to the geological structure and general topography of the landscape of a region.

Plastron The lower shell of a turtle.

Posterior Away from the head of an animal and toward the tail.

Radiotelemetry A method using a radiotransmitter attached to or implanted in an animal to track its movements by locating it using a directional antenna and radio receiver.

Scutes The modified enlarged scales or plates that form the carapace and plastron of a turtle shell.

Serrated marginals Marginal scutes that form a jagged edge.

Specialist An animal restricted in its choice of diet or habitat.

Species Typically an identifiable and distinct group of organisms capable of interbreeding and producing viable offspring under natural conditions.

Subspecies A taxonomic unit, or "race," within a species, usually defined as morphologically distinct and occupying a geographic range that does not overlap with that of other "races" of the species. Subspecies may interbreed naturally in areas of geographic contact (*see* Intergrade).

Supramarginals A narrow row of scutes on the carapace between the costal and marginal scutes that can be used to identify alligator snapping turtles.

Taxonomy The scientific field of classification and naming of organisms.

Titillation A courtship activity of some species of aquatic turtles in which the male vibrates his elongated foreclaws in the water in front of the female.

Ventral Referring to the belly or underside of an animal.

Vertebrals The central row of scutes running lengthwise down the carapace of a turtle.

Warm-blooded A nontechnical term that refers to an animal that maintains its body temperature primarily through the use of metabolic heat. *See also* Endotherm.

Further Reading

Ashton, P. S., and R. E. Ashton Jr. 2004. *The Gopher Tortoise: A Life History.* Sarasota, Fla.: Pineapple Press.

Ashton, R. E., Jr., and P. S. Ashton. 1985. *Handbook of Reptiles and Amphibians of Florida.* Part 2: *The Turtles.* Miami, Fla.: Windward Publishing.

Bartlett, R. D., and P. P. Bartlett. 1996. *Turtles and Tortoises.* Hauppauge, N.Y.: Barron's Educational Series.

Bartlett, R. D., and P. P. Bartlett. 1999. *A Field Guide to Florida Reptiles and Amphibians.* Houston: Gulf Publishing Company.

Behler, J. L., and F. W. King. 1979. *The Audubon Society Field Guide to North American Reptiles and Amphibians.* New York: Alfred A. Knopf.

Carmichael, P., and W. Williams. 2001. *Florida's Fabulous Reptiles and Amphibians.* Tampa, Fla.: World Publications.

Carr, A. 1995. *Handbook of Turtles: The Turtles of the United States, Canada, and Baja California.* 1952. Reprint. Ithaca, N.Y.: Cornell University Press.

Carroll, D. M. 1991. *The Year of the Turtle: A Natural History.* Charlotte, Vt.: Camden House.

Conant, R., and J. T. Collins. 1991. *A Field Guide to Reptiles and Amphibians of Eastern and Central North America.* 3rd ed. Boston: Houghton Mifflin.

Dodd, C. K., Jr. *North American Box Turtles: A Natural History.* Norman: University of Oklahoma Press.

Dundee, H. A., and D. A. Rossman. 1989. *The Amphibians and Reptiles of Louisiana.* Baton Rouge: Louisiana State University Press.

Ernst, C. H., and R. W. Barbour. 1989. *Turtles of the World.* Washington, D.C.: Smithsonian Institution Press.

Ernst, C. H., J. E. Lovich, and R. W. Barbour. 1994. *Turtles of the United States and Canada.* Washington, D.C.: Smithsonian Institution Press.

Gibbons, J. W. 1983. *Their Blood Runs Cold: Adventures with Reptiles and Amphibians.* Tuscaloosa: University of Alabama Press.

Gibbons, J. W. 1990. *Life History and Ecology of the Slider Turtle.* Washington, D.C.: Smithsonian Institution Press.

Gurley, R. 2003. *Keeping and Breeding Freshwater Turtles.* Ada, Okla.: Living Art Publishing.

Highfield, A. C. 1996. *Practical Encyclopedia of Keeping and Breeding Tortoises and Freshwater Turtles.* London: Carapace Press.

Klemens, M. W. 2000. *Turtle Conservation.* Washington, D.C.: Smithsonian Institution Press.

Lohoefener, R., and R. Altig. 1983. *Mississippi's Amphibians and Reptiles: A Distributional Survey.* NSTL Station, Mississippi State University.

Martof, B.S. 1956. *Amphibians and Reptiles of Georgia: A Guide.* Athens: University of Georgia Press.

Martof, B. S., W. M. Palmer, J. R. Bailey, J. R. Harrison III, and J. Dermid. 1980. *Amphibians and Reptiles of the Carolinas and Virginia*. Chapel Hill: University of North Carolina Press.

Mitchell, J. C. 1994. *The Reptiles of Virginia*. Washington, D.C.: Smithsonian Institution Press.

Mount, R. H. 1975. *The Reptiles and Amphibians of Alabama*. Auburn, Ala.: Auburn University Agricultural Experiment Station.

Palmer, W. M., and A. L. Braswell. 1995. *Reptiles of North Carolina*. Chapel Hill: University of North Carolina Press.

Pope, C. H. 1967. *Turtles of the United States and Canada*. New York: Alfred A. Knopf.

Pritchard, P. C. H. 1967. *Living Turtles of the World*. Neptune, N.J.: T. F. H. Publications.

Pritchard, P. C. H. 1979. *Encyclopedia of Turtles*. Neptune, N.J.: T. F. H. Publications.

Pritchard, P. C. H. 1989. *The Alligator Snapping Turtle: Biology and Conservation*. Milwaukee: Milwaukee Public Museum.

Ruckdeschel, C., and C. R. Shoop. 2006. *Sea Turtles of the Atlantic and Gulf Coasts of the United States*. Athens: University of Georgia Press.

Trauth, S. E., H. W. Robison, and M. V. Plummer. 2004. *The Amphibians and Reptiles of Arkansas*. Fayetteville: University of Arkansas Press.

Zappalorti, R. T. 1976. *The Amateur Zoologist's Guide to Turtles and Crocodilians*. Harrisburg, Pa.: Stackpole Books.

Zim, H. S., and H. M. Smith. 2001. *Reptiles and Amphibians*. A Golden Guide. New York: St. Martin's Press.

Acknowledgments

We wish to acknowledge our appreciation to colleagues who graciously reviewed individual species accounts and offered suggestions for improvement: Kimberly Andrews, Matt Aresco, Mark Bailey, Roger Birkhead, Russ Bodie, Vincent Burke, Justin Congdon, Ken Dodd, Mike Dorcas, Jim Godwin, Judy Greene, Craig Guyer, Jim Harding, Brian Horne, John Iverson, Dale Jackson, John Jensen, Peter Lindeman, Jackie Litzgus, Jeff Lovich, Ken Marion, Peter Meylan, Tony Mills, Joe Mitchell, Paul Moler, Dave Nelson, Mike Plummer, Mike Seidel, Steve Shively, Stan Trauth, Dawn Wilson, Bob Zappalorti. For providing expertise and assistance in creating the species distribution maps, we thank Tom Akre, John Iverson, and Deno Karapatakis. Peri Mason generously provided original artwork. Two anonymous reviewers read the entire manuscript and provided many helpful suggestions. We are grateful to Margaret Wead, who helped scan slides, and to Teresa Carroll and Juanita Blocker, who also provided assistance during the preparation of this manuscript.

Credits

The authors would like to thank the following individuals and organizations for providing photographs:

Matthew J. Aresco
Photographs on pages 102 (top), 110 (top), 112 (top), 116 (left), 117 (top), 120–121, 151 (right), and 231 (left).

R. D. Bartlett
Photographs on pages 8, 15 (left), 52 (both), 57 (right top and middle), 60 (right), 63 (top), 71 (top), 77 (top left), 80 (bottom left), 82, 84 (both), 90 (top), 110 (bottom), 115 (top), 126 (top), 129, 132 (left), 133 (bottom right), 140 (top right and bottom left), 142, 143 (bottom), 145 (bottom), 148, 150 (bottom), 153 (second), 153 (second from bottom), 175 (both), 183 (top), and 205 (bottom).

Will D. Brown
Photograph on page 10.

Kurt A. Buhlmann
Photographs on pages 9 (both), 14, 19 (top left and right), 20 (top right and bottom), 23 (bottom), 24 (both), 26 (bottom), 27 (bottom left and right), 28 (left), 29 (bottom), 32 (top left), 32 (top right), 32–33 (bottom), 33 (top left and right), 34 (both), 35 (all), 36, 37 (bottom), 38 (top), 38–39 (middle), 48–49 (bottom), 49 (right), 50 (top left), 55, 59 (bottom), 64 (top), 64–65 (bottom), 66 (both), 67 (bottom), 69 (top), 72 (bottom left), 77 (right), 78 (top), 80 (bottom right), 85 (both), 86 (both), 87 (top right), 88, 92, 97 (right), 99 (bottom left and right), 101, 104, 105 (bottom left and right), 107 (both), 115 (bottom left and right), 116 (right), 118 (bottom), 123 (both), 124 (both), 130 (both), 133 (top right), 139 (bottom left and right), 141 (left), 151 (left), 154 (top), 157 (left), 165 (top), 166 (left and right), 177 (male), 183 (bottom left), 190 (top), 195 (left), 203, 204 (both), 206 (bottom left), 207 (both), 211 (bottom center), 212 (bottom three), 214 (middle and bottom), 215 (both), 219, 221 (top), 224 (right), 226, 229 (right top and middle), 232 (both), and 233.

Tony Campbell/istockphoto.com
Photograph on page ii.

John L. Carr
Photographs on pages 19 (middle right) and 179 (both).

Dave Collins
Photographs on pages 60 (left), 67 (top), 72 (right), 94 (top), 185 (both), 193, 197, and 229 (left).

C. Kenneth Dodd
Photograph on page 19 (middle left).

Michael E. Dorcas
Photograph on page 50 (bottom right, with Tracey Tuberville).

Andrew M. Durso
Photograph on page 13.

Saul Friess
Photograph on page 76 (middle).

Michael Gibbons
Photograph on page 225.

Whit Gibbons
Photographs on pages 27 (top), 76 (bottom), 222 (both), and 223 (left).

Andrea Gingerich/istockphoto.com
Photograph on pages 200–201.

James Godwin
Photographs on pages 6, 12, 21 (bottom), 54, 76 (top), 96–97 (top), 114, 125, 126 (bottom), 139 (top right), 140 (bottom right), 146 (both), 155 (right), 158 (all), 159, 160 (top), 161 (both), 162 (both), 169 (top), 182 (top), and 184.

Kedar Gore
Photograph on page 217.

Sylvia Greenwald
Photograph on page 78 (bottom).

Cris Hagen
Photographs on pages 3, 31, 39 (both), 75, 81 (both), 113 (top), 132 (right), 155 (left), 160 (bottom), 163, 167 (top), 169 (bottom), 172, 178 (top), 188 (top), 191, 194 (top center and bottom), 195 (center), 196 (all except right second from bottom), 198 (bottom), 206 (top), 211 (top left and right and bottom left), and 216 (bottom).

James H. Harding
Photographs on pages 57 (left), 63 (bottom), 71 (bottom), 91 (top), 99 (top), 173 (left), and 177 (nesting female).

Brian D. Horne
Photographs on pages 15 (right), 165 (bottom), 167 (bottom), and 186 (both).

John B. Iverson
Photographs on pages i, 18 (top), 21 (top), 28 (right), 44, 48 (top), 51, 53, 58 (bottom), 79, 80 (top), 103, 119, 140 (top left), 149 (middle right), 171 (top), 176, 177 (plastron), 182 (bottom), 189 (top), 211 (bottom right), 213 (all), 214 (top), and 216 (middle).

Robert L. Jones
Photographs on pages 187 and 190 (bottom).

Trip Lamb
Photographs on pages 2 (bottom), 50 (bottom left), 93 (top), and 212 (top).

Peter V. Lindeman
Photograph on page 144 (top).

Jeff Lovich
Photograph on page 199.

Thomas M. Luhring
Photographs on pages 11 (top), 22 (both), 58 (top), 93 (bottom left and right), 95, 131, 134, 149 (right bottom and left top and bottom), 149 (bottom left), 153 (bottom), 156 (bottom), 178 (bottom), 180 (both), 181 (bottom), 183 (bottom right), 188 (bottom), 189 (right), and 216 (top).

Michael Marchand
Photograph on page 26 (top).

Sherry Melancon
Photographs on pages 135 (top), 136, and 137.

Joseph C. Mitchell
Photograph on page 166 (center).

Nancy Nehring/istockphoto.com
Photograph on page viii.

David Nelson
Photographs on pages 145 (top) and 147.

Charles Scott Pfaff
Photograph on page 228 (top).

Malcolm Pierson
Photographs on pages 25 (top), 37 (top three), 38 (bottom), and 100.

Michael V. Plummer
Photographs on pages 127, 141 (right), 144 (bottom), 168 (top), 173 (right), and 174.

Rainer Schmittchen/istockphoto.com
Photograph on page 1.

David E. Scott
Photographs on pages 11 (right), 150 (top), and 221 (bottom left).

Richard A. Seigel
Photograph on page 194 (top right).

Savannah River Ecology Laboratory
Photographs on pages 29 (top), 30, 90 (bottom), 109 (right), 198 (top), 205 (top), 218, and 223 (right two).

Todd Stailey
Photographs on pages 4, 41, 89, 117 (bottom), 181 (top), and 230.

Sean Sterrett
Photographs on pages 152 and 154 (bottom).

Hollis Ann Stewart
Photograph on page 231 (right).

Tracey Tuberville
Photographs on pages 18 (bottom), 50 (top right), and 221 (bottom right).

Robert Wayne Van Devender
Photographs on pages 2, 23 (top), 47 (top), 57 (bottom right), 59 (top), 65 (right two), 69 (bottom), 70, 77 (bottom left), 83 (both), 86–87 (bottom), 91 (bottom), 96 (bottom), 118 (top), 128, 135 (bottom), 138, 139 (top left), 143 (top), 153 (middle), 162 (middle), 164, 170 (both), 194 (top left), and 195 (right).

Michael R. Vaughan
Photograph on page 220.

Rico Walder
Photograph on page 56.

Pai Wei/istockphoto.com
Photograph on pages 208–209.

Lucas Wilkinson
Photographs on pages vii and 19 (bottom).

J. D. Willson
Photographs on pages 16, 98, 102 (bottom), 112 (bottom), 133 (left), 149 (top right), and 196 (right second from bottom).

Zappalorti, Robert T.
Photographs on pages v, 7, 17, 20 (top left), 25 (bottom), 40, 47 (bottom), 61, 72 (top left), 74 (both), 94 (bottom), 105 (top), 106 (both), 108 (top), 108–109 (bottom), 111, 113 (bottom), 153 (top), 156 (top), 157 (right), 168 (bottom), 171 (bottom), 206 (bottom right), 228 (bottom), and 229 (bottom right).

Index of Scientific Names

Boldface page numbers refer to species accounts. *Italicized* page numbers refer to illustrations.

Index of Common Names

Boldface page numbers refer to species accounts. *Italicized* page numbers refer to illustrations.